Hunting Marfa Lights

by
James Bunnell

Lacey Publishing Company
Benbrook, Texas

Library of Congress Control Number: 2009903408

ISBN-13: 9780970924940

Printed and bound in the United States of American

Lacey Publishing Company
Benbrook, TX 76132-1003
www.marfalightsresearch.com

To
Kerr and Mary Belle Mitchell

Photograph from one of the Mitchell's security cameras

Other books by this author

Seeing Marfa Lights

Night Orbs

Table of Contents

Please Note

Figures presented in this edition are reproduced in black,
white and grey tones. Some of these were originally taken in color.
Visit www.marfalightsresearch.com to see figures in color.

Acknowledgments

I want to extend very special thanks to Mitchell Flat ranchers who have generously accommodated my monitoring station equipment and assisted me at every turn of this complex project.

I am especially grateful to William Kerr and Mary Belle Mitchell, whose steady support of my research effort has made this long-term effort possible. I also want to thank my brother Will for his encouragement, comments, electrical knowledge and contributions of helpful assistance throughout this investigation.

The Marfa Lights story would not be complete without the accounts of all the people who have published or otherwise shared their Marfa Lights stories, including the 1945 account published by San Angelo Standard Times and the papers published by Elwood Wright and Pat Kenney, Edson Hendricks, and Sharon Eby Cornet. I have been the very fortunate recipient of personal accounts from Fred Tenny, Van, Alton Sutter, Dirk and Sarah Vander Zee, Linda

Lorenzetti, Bill Jones, Dan and Wife, James Nixon III, M. Bennett and Spouse, Linda Armstrong, Don Batory and family, Rob Grotty, and Linda Quiroz. These stories presented in Appendix A are invaluable contributions that reveal the depth and complexity of Marfa Light experiences.

My special thanks are offered to Dr. Karl Stephan, Dr. Irwin Wieder, and Dr. Sten Odenwald for their interest and consultations regarding various technical issues. I am especially grateful to Karl Stephan for his help with spectrometer and spectral-related questions and issues.

I am also indebted to Dr. Judith Brueske and Charlotte Allen for their interest and support, and to Kevin Webb for his help with moving Snoopy.

Last, but not least, my deep gratitude and appreciation are extended to my wife, Sandra Martin Dees, for her unflagging support of this investigation, her skillful and most helpful editing of this manuscript, and for the many Marfa days and nights she has devoted to helping me in my hunt for mystery lights.

James Bunnell, 2009

"...he turned his head to the right and could clearly see this odd light through the untinted back window of his truck. It was a ball of green light that did not radiate or show any evidence of a light beam. Rob was driving 65 mph and this strange light seemed to be about 5 mph faster because it was gaining on him." -- From Story 32

Introduction

Mysterious lights are reported in many locations worldwide, but one of the best known of these sites is near the small West Texas town of Marfa. In fact, the Marfa Lights are so well known that the State of Texas has created a roadside park for travelers who wish to stop by and take a look. Marfa is slowly gaining worldwide recognition as the home of mischievous nocturnal lights that shine, pulse, dance, and do amazing things to delight lucky observers. These nocturnal happenings are observed mostly east of town in a region known as Mitchell Flat. The question is, of course, are Marfa Lights real and mysterious, or simply folklore hyped by locals to attract tourist dollars?

If you talk to people who have been there, you will find a wide range of answers. At one extreme are the skeptics who say they saw only car lights traveling a distant mountain road; the only mystery to these folks is why so many people stare excitedly into the night at those "car lights." At the other extreme are people who claim they not only saw mysterious lights, but were absolutely stunned by the experience. In between those extremes are the majority of people who went to the View Park, looked into the night and saw lights, but left unsure as to what they had actually seen.

So, the reader may wonder, "What is going on? What is this fuss all about?" Based on my investigation of Marfa Lights phenomena, I believe that in most instances, mysterious lights seen from the Marfa Lights View Park do have explainable sources. Some are indeed motor vehicle lights. Others come from many explainable light sources, as discussed in Part II of this book.

After eight years of investigation, I am equally convinced that there are also mysterious lights that are the real foundation behind all of the fuss over Marfa Lights. They are indeed rare, but real and most unusual. Stories of people who have experienced these marvelous displays speak volumes. Readers will find a generous sample of these stories in Appendix A. Part I of this book contains photographs and descriptions of mystery light behaviors, and chronicles my evolving investigation into these complex phenomena and growing knowledge about them.

As my investigation has proceeded, some aspects of these mysteries have come into focus while other aspects have only grown more puzzling. Some observers have suggested these phenomena are UFOs, but I do not share that view. Others have suggested that Marfa Lights are paranormal in nature, but I do not subscribe to that concept either. It is my belief that some of the mysterious lights seen near Marfa, and many other locations worldwide, constitute natural phenomena not yet fully studied or understood by our scientific community.

In the 1960s, a Sul Ross University physics professor mounted an effort to look into the source of mystery lights and concluded that people were seeing car lights and ranch lights. In May 2004, the University of Texas Society of Physics Students conducted a four-night investigation using traffic volume-monitoring equipment, video cameras, binoculars, and chase cars. They concluded that all observed "mysterious-looking" lights could be attributed to vehicle traffic on US 67 (the road from Presidio that descends out of mountains on its way to Marfa and is visible from the Marfa Lights View Park). Other more limited studies and amateurish YouTube videos have only served to further increase the skepticism of many people who have not experienced the phenomena.

Nevertheless, these mysterious lights continue to appear. The difficulty is that true instances of Marfa Lights -- those that are not vehicle lights or ranch lights -- are rare and unpredictable events. I believe that steady voices of rational witnesses and contributions of long-term investigations, such as the one reported in this book, will eventually be heard above cynical critics and cause an awakening of scientific interest, because nothing is more exciting to science (or to most of us for that matter) than encounters with the unknown -- natural events that we do not yet understand. Perhaps this book will contribute a small step toward that awakening. In any case, hunting Marfa Lights, as readers are about to discover, is fun, exciting, and hard work.

"When the light was first turned on it was dim orange. As it swung downward it became a bright white. At the turnoff position it dimmed to nothing." -- From Story 31

To actually witness these mysterious lights does require luck or patience because this phenomenon is indeed rare, but it is one well worth waiting to see. If you do visit Marfa, be sure to stop at the Marfa Lights View Park east of town. If luck is with you, as it has been with me, you just might see mysterious orbs of light suddenly appear above desert foliage. These balls of light may remain stationary as they pulse on and off with intensity varying from dim to almost blinding brilliance. Then again, these ghostly lights may dart across the desert in complete defiance of prevailing winds, or perform splits and mergers: First you see one light, and then another emerges out of the first to move left or right. The "offspring" is usually less intense but frequently more dynamic. You may see one or more offspring lights oscillating back and forth as if passing through, behind, or orbiting the parent light. Offspring lights will sometimes merge back into the parent, but multiple light splits can also occur, resulting in a dazzling display of bobbing, pulsing, dynamic and ever-mysterious lights.

Light colors are usually yellow-orange but other hues, including green, blue and red are also seen. Red is the most frequent alternative color. I have not witnessed a mystery light that was red from start to finish, but other people have. Marfa Lights are typically yellow-orange or orange for most of their lifetimes and will, on occasion, convert suddenly and completely into brilliant red. The red state is typically of short duration after which the light either goes out completely or may flash once more into brilliant yellow-orange. Transitions into blue or green do occur but are

14

rarely observed.

Marfa is a high-desert basin surrounded by beautiful mountains and mesas. Marfa Mystery Lights (MLs) usually fly above desert vegetation but below background mesas, with altitudes varying anywhere from 400 feet to as low as 2 or 3 feet above ground level. There are exceptions. One type of mystery light flies high in the night sky entirely above local horizons. High flying MLs are typically fast moving and more dynamic. Another type of mysterious display is more "curtain-like" in appearance, containing a rainbow array of colors.

Over time, legends have developed around the Marfa Lights but, fortunately, they have never acquired a reputation for being harmful. There are tales of Marfa Lights relentlessly following automobiles and giving fleeing drivers a good scare when they are unable to outrun the lights no matter how fast they might drive (see Stories 32-34 in Appendix A). There have also been stories of Marfa Lights entering parked automobiles, frightening occupants with a momentary flash of brilliant light, but leaving them otherwise unharmed.

One legend even credits Marfa Lights with beneficence. According to this tale, a college professor became lost while hiking in the desert near Marfa. Since the elevation is almost a mile high, Marfa nights can be bone-chilling cold even in warm weather, and temperatures drop quickly after the sun goes down. Within half an hour after sunset, the lightly-dressed professor, shivering and badly scared, spotted a light. Believing it to be a ranch light, he headed in that direction in hopes of finding shelter. To his amazement, the mysterious light lead him to his truck. As he reached the safety of his vehicle, the benign light winked out. True or not, I like this story because it runs counter to our natural inclination to fear the unknown. And there is no better one-word descriptor for Marfa Lights than the word "Unknown."

As mysterious and intriguing as Marfa Lights may be, I am convinced that they are natural events well deserving of scientific study and attention. It has been my pleasure and privilege to investigate these elusive phenomena over the last few years. This book is the story of that investigation.

"The light was large and moving rapidly in her direction. At first she thought it must be a large truck wanting to pass, but as it came closer she could see it was a single bright light. As collision seemed imminent, she involuntarily pressed the accelerator to the floor but could not outrun the light." -- from Story 34

Mitchell Flat

Marfa Mystery Lights (I call them MLs, a short for mystery lights) are most often seen in Mitchell Flat, located approximately nine miles east of Marfa and about 80 miles northeast of the Big Bend National Park.

Similar mystery lights are seen in many locations, not only in this country (e.g., Brown Mountain Lights, North Carolina; Gurdon Light, Gurdon, Arkansas; Hebron Light, Hebron, Maryland; Joplin Spook Light, Northeast Oklahoma; Anson Lights, Abilene, Texas; Bingham Light, Dillon, South Carolina; Chapel Hill Light, Chapel Hill, Tennessee; Codgell Spook Light, Codgell, Georgia) but in other countries as well (e.g., Dovedale Light, Dovedale, UK; Ontario Lights, Ontario, Canada; Min Min Lights, Min Min,

Australia; St. Albans Light, England; Hessdalen Valley Lights, Hessdalen, Norway[1]). A global study of worldwide mystery lights will need to wait for another day. My study, and the subject of this book, is limited to light phenomena observed in the vicinity of Marfa, Texas.

Marfa is an excellent place to study unusual light phenomena because it is a region of high desert that has wide and distant vistas, permitting uncluttered observation, both near and far. We live in an age and a country that loves light and Marfa is no exception. Artificial light sources in the form of ranch lights, vehicle lights, and aircraft lights abound, but these light sources can be readily identified and distinguished from Mystery Lights. Methods for doing this are discussed in Appendix B.

First let me give you some background on the town of Marfa and then a few observations on the unique geology of the region.

Marfa

Marfa is a community located in West Texas about 60 miles south of Interstate Highway 10 at the junction of US highways 67 and 90. For anyone wishing to enter coordinates into their navigation computer, Marfa is located at N30 degrees 18.3 minutes; W104 degrees 1.5 minutes. The town started life in the 1800s as a water stop for the Galveston, Harrisburg and San Antonio Railroad. It grew quickly and became the county seat.

During the Second World War, good flying weather made Marfa an attractive location for an advanced flying school. In February of 1942, a decision was made to let a two-million dollar contract for development of an Army airfield to be used for basic multi-engine training. The Marfa Army Airfield became operational in December 1942 on the one-year anniversary of Pearl

Harbor. The Army Airfield was decommissioned following the war and the land was returned once again to ranching purposes.

Today Marfa is a small ranching and art community with a population of around 2100 people. The largest employer is the US Border Patrol with Marfa being Headquarters for the regional sector. Marfa has been building a reputation as a good location for movie making (e.g., *Giant* with Rock Hudson, Elizabeth Taylor, and James Dean, *There Will Be Blood* with Daniel Day Lewis, and *No Country for Old Men* with Javier Bardem and Tommy Lee Jones).

But Marfa's most enduring claim to fame is surely the mystery lights that frolic in the cold night desert air east of town in Mitchell Flat. You might wonder, if the Army Airfield was located in Mitchell Flat (where MLs are most often seen), did Army personnel see them and, if so, what did they think they were? The answer is that Army personnel absolutely did see MLs on many nights and made multiple attempts to track them down.

If you are a pilot and see mysterious lights in the night, doesn't it seem logical to try and chase down MLs from the air? Some aircrews felt that it did, and unauthorized attempts were certainly made. The most enduring story from that time frame is about flour bombing.[2] These were, after all, pilots training to drop bombs, right? The scheme (some would say hair-brained, but what do they know?) was simple enough. Be ready to fly and when an ML is reported by people on the ground, take off, fly over the area, locate the ML, swoop down low and bomb it with bags of flour. The notion was that flour bags would break on impact causing a burst of flour to mark the spot. The marked spot could then be found the next day in daylight, enabling a detailed search for any possible source of the mystery light.

Were any of these nightly bombing raids ever carried out? We do not know for sure but I suspect they did happen. The method was not entirely without hazard. On a moonlit night a pilot with night vision intact might be able to execute such a mission with some bombing error and still return safely to the field. But MLs do not always show when you might like them to appear. On a dark night or if the pilot has lost his night vision, such a mission could be downright risky at best, given the undulating hilly terrain with potential for unexpected ground contact. In any case, the technique did not produce any answers. Even if the flour spot could be found there would be nothing left behind to reveal the origin of the lights. I say this based on modern techniques that have made it possible for me to visit ML sites during daylight. I will explain how that was accomplished later but should note for now that I have never found any surface clues and I have had opportunities to search multiple sites.

There is an interesting story about how MLs almost caused an aircraft to crash.[3] The air training syllabus required night landings at some of the auxiliary fields. A student pilot mistook a line of MLs for auxiliary airfield landing lights and approached dangerously close to the ground before recognizing that those MLs were not runway lights. Realizing his mistake, he applied full power and pulled up in time to avert disaster.

Geographical Uniqueness

Given the unusual nature of mystery lights it is logical to ask if the region is geologically unique in some way that might account for these phenomena. Because Marfa is almost a mile high, temperatures plunge dramatically with onset of darkness, sometimes dropping 30 or 40 degrees within the first couple of hours. Strong

20

wind currents are common and wind shear can be seen in cloud formations that are traveling rapidly past each other in opposite directions. Strong turbulent winds, mountain ranges, and high desert sometimes combine to create conditions ideal for temperature inversions and that fact might be an important factor contributing to ML reports as will be discussed in Part II.

In addition to temperature oddities, the region also bears the marks of the intense volcanic activity that occurred thirty-one million years ago. Paisano and Chinati mountains were then active volcanoes spewing out great quantities of volcanic ash that completely covered Mitchell Flat with a thick layer of compressed ash known as tuff.[4] Today a fairly thin layer of top soil sits on top of the tuff and supports desert plant life. Bare tuff can readily be seen on ranch roads in the area (see Figure 1). Ash was also spewed into inland seas in the region; volcanic ash falling into salt water forms a type of crystal known as zeolites. Zeolites are so common in the region that they are being commercially mined near Casa Piedra (30 miles south of Mitchell Flat) where they are sufficiently abundant to be scooped up with a front loader.

Figure 1. Geologically, Mitchell Flat is indeed unusual because ancient volcanoes deposited a thick layer of compressed volcanic ash known as tuff. The tuff can be as much as 50 to 100 feet thick and it covers all of Mitchell Flat and goes up the sides of surrounding mountains. Ranch roads in the area are often graded down to tuff revealing how thin the top layer of soil actually is.

"A group of lights flew over the roadway near their location. Most of the lights continued on course but one light slowed, causing Sutter's son to announce, 'Look Dad! One of them is landing.'"
-- from Story 6

Marfa Lights View Park

From Marfa, drive east on U.S. Highway 90/67 approximately 9 miles. The Marfa Lights View Park is located on your right (south side of the roadway). From Alpine, drive west on U.S. Highway 90/67 approximately 17 miles. The Park will be on your left. For clarity, in this book I will refer to the Marfa-Alpine highway as "US 90" and the highway from Presidio to Marfa as "Highway 67."

A blue and white sign identifies the site with an arrow pointing south to the park. Below the arrow the sign reads, "Nighttime only." In my guidebook, *Seeing Marfa Lights*, I joked that this

concisely captures in three words all of
what had been learned about the lights. I
know much more now -- you will read it
in this book.

The Marfa Lights View Park is
ideally situated for seeing mystery light
phenomena. It is also located right next to the site of the World War
II Army Airfield, although the many hangars and Army buildings
have long since been demolished and are now represented only by
concrete foundations. Remains of the adobe entrance walls to the
airfield are located just east of the Park.

Parking space is sufficient to accommodate many vehicles
including recreational vehicles (RVs) and large trucks. The drive-
through arrangement is convenient for travelers with large RVs or
trailers since backing your vehicle can usually be avoided. On a
typical evening, expect to see around ten to twenty vehicles,
including two or three RVs. As you would expect, warm weather
weekends bring the largest crowds.

In 2003, the View Park was significantly expanded by Texas
Department of Transportation to include a view shelter, rest rooms,
and informational plaques connected by park trails. On the following pages are a series of photographs and maps that provide a
graphical sketch of this unusual facility.

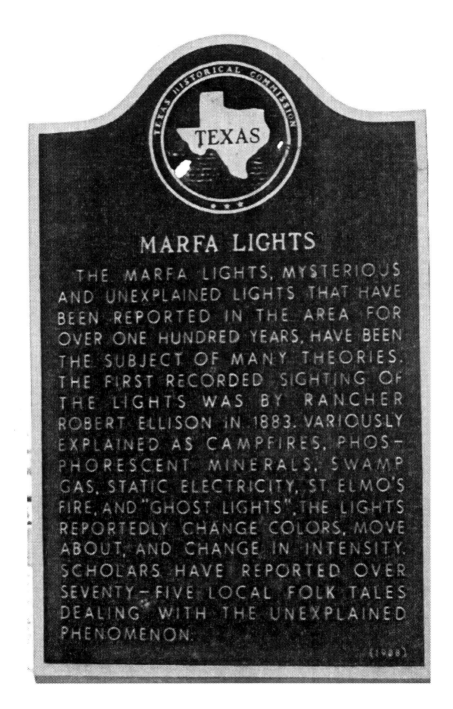

Figure 2. This Texas Historic Marker briefly summarizes what Marfa Lights are all about.

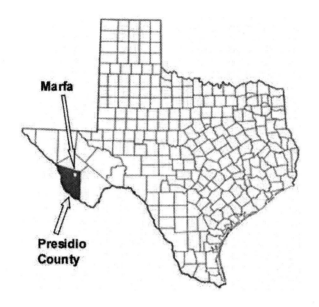

Map 1. Marfa is the county seat for Presidio County in West Texas.

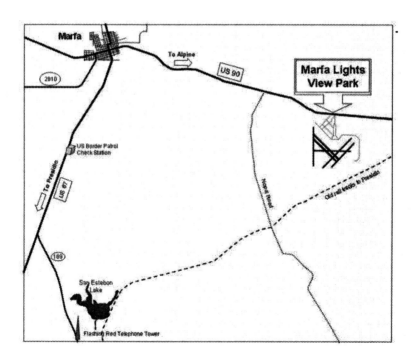

Map 2. The Marfa Lights View Park (MLVP) is nine miles east of Marfa on US 90 next to where an Army Airfield was located during WWII.

26

Figure 3. Driving from Marfa or Alpine you will see an advisory sign one mile before reaching the View Park.

Figure 4. Marfa Lights View Park provides both a convenient shelter for viewing the lights and a rest stop for travelers.

Figures 5 & 6. The View Shelter provides an elevated concrete deck for viewing and photography.

28

Figure 7. The plaque directly in front of the View Shelter explains that Apache Indians believed mystery lights to be fallen stars.

Figure 8. There are eight informational plaques that can be reached by following gravel trails.

29

Marfa Lights View Park (MLVP) and Shelter

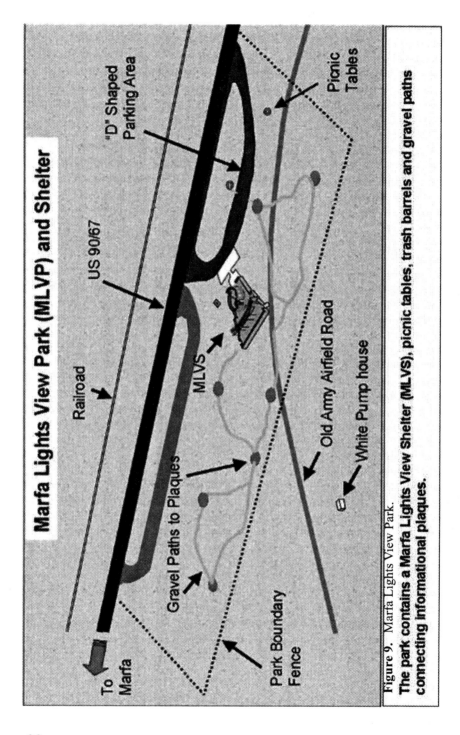

Figure 9. Marfa Lights View Park.
The park contains a Marfa Lights View Shelter (MLVS), picnic tables, trash barrels and gravel paths connecting informational plaques.

30

Figure 10. Walking Park trails before dark.

Figure 11. Visitors use the center telescope while waiting for darkness to fall at the Marfa Lights View Park.

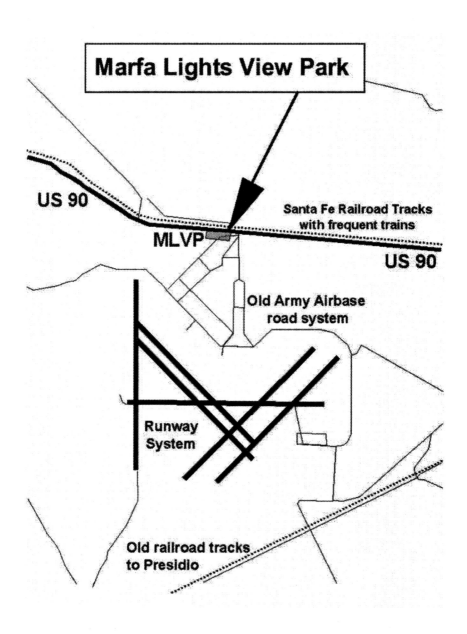

Map 3. The Marfa Lights View Park (MLVP) is located near what remains of an Army Airfield that existed during WWII to train bomber pilots. Only the runways and building foundations remain. Some of the foundations can be seen from the View Park before dark, but the runways are too low to be visible.

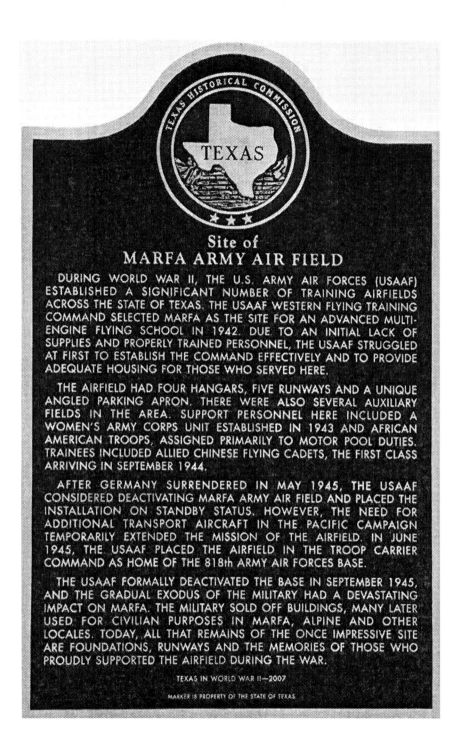

Site of
MARFA ARMY AIR FIELD

DURING WORLD WAR II, THE U.S. ARMY AIR FORCES (USAAF) ESTABLISHED A SIGNIFICANT NUMBER OF TRAINING AIRFIELDS ACROSS THE STATE OF TEXAS. THE USAAF WESTERN FLYING TRAINING COMMAND SELECTED MARFA AS THE SITE FOR AN ADVANCED MULTI-ENGINE FLYING SCHOOL IN 1942. DUE TO AN INITIAL LACK OF SUPPLIES AND PROPERLY TRAINED PERSONNEL, THE USAAF STRUGGLED AT FIRST TO ESTABLISH THE COMMAND EFFECTIVELY AND TO PROVIDE ADEQUATE HOUSING FOR THOSE WHO SERVED HERE.

THE AIRFIELD HAD FOUR HANGARS, FIVE RUNWAYS AND A UNIQUE ANGLED PARKING APRON. THERE WERE ALSO SEVERAL AUXILIARY FIELDS IN THE AREA. SUPPORT PERSONNEL HERE INCLUDED A WOMEN'S ARMY CORPS UNIT ESTABLISHED IN 1943 AND AFRICAN AMERICAN TROOPS, ASSIGNED PRIMARILY TO MOTOR POOL DUTIES. TRAINEES INCLUDED ALLIED CHINESE FLYING CADETS, THE FIRST CLASS ARRIVING IN SEPTEMBER 1944.

AFTER GERMANY SURRENDERED IN MAY 1945, THE USAAF CONSIDERED DEACTIVATING MARFA ARMY AIR FIELD AND PLACED THE INSTALLATION ON STANDBY STATUS. HOWEVER, THE NEED FOR ADDITIONAL TRANSPORT AIRCRAFT IN THE PACIFIC CAMPAIGN TEMPORARILY EXTENDED THE MISSION OF THE AIRFIELD. IN JUNE 1945, THE USAAF PLACED THE AIRFIELD IN THE TROOP CARRIER COMMAND AS HOME OF THE 818th ARMY AIR FORCES BASE.

THE USAAF FORMALLY DEACTIVATED THE BASE IN SEPTEMBER 1945, AND THE GRADUAL EXODUS OF THE MILITARY HAD A DEVASTATING IMPACT ON MARFA. THE MILITARY SOLD OFF BUILDINGS, MANY LATER USED FOR CIVILIAN PURPOSES IN MARFA, ALPINE AND OTHER LOCALES. TODAY, ALL THAT REMAINS OF THE ONCE IMPRESSIVE SITE ARE FOUNDATIONS, RUNWAYS AND THE MEMORIES OF THOSE WHO PROUDLY SUPPORTED THE AIRFIELD DURING THE WAR.

TEXAS IN WORLD WAR II—2007

MARKER IS PROPERTY OF THE STATE OF TEXAS

Figure 12. Marfa Army Airfield plaque.

Figure 13. Entrance gate to what once was the Army Airfield is visible looking northeast from the Marfa Lights View Park.

Figure 14. This plaque is attached to the right wall of what used to be the entrance to a grand Airfield.

34

Figure 15. Marfa Army Airfield was used during WWII to train bomber pilots.

Figure 16. Before dark it is possible to see many of the remaining foundations of the Airfield.

Part I
Hunting Light

"At that moment two Mystery Lights were observed moving southwest to northeast very fast. Kenney had the impression that they were traveling 150 to 200 m.p.h. The first light crossed the road approximately a thousand feet away and continued east where it seemed to meet and merge with a third light. The second light crossed the road close to where the first light had crossed but the two geologists had moved closer by then and were only about 200 feet away when the ML paused and hovered, pulsing on and off approximately three feet above the road in front of them." --
from Story 2

Critical Events

A Grandfather's Gift

When I am giving presentations on Marfa Lights one of the questions that I sometimes get is, "When did you first learn about the Marfa Lights?" Phrased that way, the question is sure to invoke memories from childhood including memories of the man I was named after, my grandfather, James Henry Fortner. My parents divorced when I was age three and Mother moved my brother Will and me back to Marfa to live with her parents. Father, a U.S.

Customs agent in Presidio, relocated to the El Paso port of entry and his visits to us were usually limited to once a year. Grandfather Fortner became my male role model. My grandfather was county clerk for Presidio County but to me he was much more. A kind and gentle man, college educated as a pharmacist, he was admired and respected by those who knew him. He had a keen and inquisitive intellect with many interests including medicine, ranching, farming, and astronomy. I still remember his taking me out in the yard in Marfa to show me a comet even though I was too young to remember much about it.

It was Grandfather Fortner who first took me to see the Marfa Lights. I was very young at the time. I remember the trip east of town on a beautiful night and remember seeing lights, but the experience was too long ago for me to recall details. Brother Will, who is older and whose memory of our time in Marfa is more vivid than mine, says our grandfather used a surveyor's level to measure azimuth angles of the lights we observed and was able to show that they were located in the direction of the Presidio-Marfa Highway (US 67). He believed that it was these car headlights that people were seeing and calling Marfa Lights.

Mother remarried when I was ten years old and we moved back to Presidio, a small town located on the Mexican Border, 60 miles south of Marfa. It was as a teenager in Presidio High School that I developed my first plan for studying the Marfa Lights. I studied trigonometry, physics and math and was impressed with the power of triangulation to compute distant locations. The surveyor's level that my grandfather had used to measure angles when he took us to see the Marfa Lights was owned by my uncle, John H. Fortner. Uncle John, who was Superintendent of Schools in Presidio, had acquired the instrument in connection with an earlier

40

occupation at the quicksilver mines in Terlingua. This surveyor's instrument consisted of a brass sight glass that could be rotated above a marked base to precisely measure angles. I saw in that instrument a possible way to solve the Marfa Lights mystery. I would borrow my uncle's surveyor's level, drive to Marfa, take bearings from three widely separated locations along US 90 and determine exactly where the light was located. The next day, in daylight, I would investigate the computed location and find whatever was causing the light. It seemed like a foolproof plan.

There was one small problem. Well, actually there were a number of problems, but it only took one for my plan to fail. No plan can work unless it is implemented! I was too busy with life to spend time chasing something as ordinary (they were considered ordinary in Marfa) as Marfa's ghost lights. I did find my way back to Marfa to graduate from Marfa High School, but did not venture out to look at Marfa Lights even once that I can remember during my senior year. High school was followed by university and a career in engineering that transported me far away from Presidio County, Marfa, and the Marfa Lights. I never completely forgot my desire to chase down that mystery and solve it, but my attention was on other things.

Sometime in the mid-1980s, my brother Will, also a career engineer, paid a visit to Marfa. The lights had always intrigued Will and, while there, he went out for another look armed with a compass and an aeronautical map. After dark he could see a mysterious, extremely dim green light that seemed very far away. He obtained its magnetic bearing and drew a line on his map representing the direction he had measured. The line intersected part of the train tracks to Presidio where a railroad section light was located. Ah ha! He knew that railroad section lights are green, yellow, or red and wondered if all the fuss about Marfa Lights was actually nothing more than railroad section lights. The green light he had observed remained motionless and there were no trains in sight on the Presidio tracks that night, so his guess seemed at least a possibility. Will shared his suspicion that section lights might be the light source behind ML reports and I was pleased with the thought that he might have found the answer. It is always comforting and stabilizing (for engineers at least) to move unknowns into the known column.

We both discussed his interesting discovery with our Uncle John who knew something about the Marfa Lights. It turned out that he had accomplished triangulation of the lights before I even thought about doing so. In fact, Uncle John said that he knew what they were but always stopped short of sharing his theory. After listening to Will's hypothesis that Marfa Lights might be nothing more than section lights on the Presidio rail line, Uncle John smiled and commented that Marfa Lights frequently move cross-country – something railroad section lights don't do. Hmmm, well, so much for that idea. It would not be the last time a promising idea about the Marfa Lights dissolved.

Many years later, in February 1993, our Uncle John passed away. Will and I met in El Paso to close out his small estate. One of the items we found was his surveyor's level. That instrument sparked thoughts of making a trip to Marfa to carry out my childhood triangulation scheme, but the feeling soon passed when I remembered how busy I was at work. There was no time for such indulgence. Will had gone to college in El Paso and his school had a nice museum devoted to history of the Southwest. We decided to give them the instrument and, so far as I know, that is where it is today.

First Amazing Look

My first significant encounter with the Marfa Lights would take place around Thanksgiving in the year 2000, after I had retired and moved back to Texas.

> *"Soon after dark we saw two strange lights on a compass-bearing almost due south from our viewing location. These lights pulsed independently and seemed to follow a randomly timed sequence that, in most cases, went from dark to relatively dim, flared to a higher level of brightness, then dimmed and eventually went out. Sometimes both lights would be on at the same time. The lights were orange-white although the one on the left did turn orange-red for about two cycles."* -- from Story 10

For me, the encounter (Story 10) was a life changing event. I was stunned by what we saw. The light displays on those two nights in November were stationary and could not have been car lights — they were more than unusual. If they were not mirages or something that could be explained, it meant that Marfa Lights were phenomena existing outside our domain of understood events. Wouldn't that be something! Considering their potential significance, it seemed ironic that I had grown up in the same county where this remarkable prize still waited patiently to be discovered.

My initial thought was that someone had already found a clear and logical explanation. I knew many smart people had investigated Marfa Lights in the past. Surely someone had broken the code. All I needed to do was minor research to find answers that must be published somewhere. I visited the Marfa Book Company (an unexpected treasure of a bookstore on Main Street) and purchased what they had on the subject, including **The Marfa Lights,** a collection of firsthand accounts, with comments and discussion by Dr. Judith M. Brueske. Dr. Brueske's booklet has many interesting ML stories and includes discussion of various ideas that have been suggested by people over the years in an effort to explain these lights. The stories increased my interest but I could see problems with all the suggested light sources. No one had been able to advance a convincing explanation.

On the drive home I looked forward to finding a better answer using the power of the Internet. As expected, the Internet was loaded with many Marfa Lights discussions and ideas that varied from naked skepticism to unsupported ghostly concepts. I easily found fault with every idea presented. I was amazed to realize that this long-running mystery seemed to be still unsolved. No one was able to offer a convincing explanation; all explanations were flawed. That meant the door remained open for me to seek my own answers.

Strategy

Before beginning my quest to investigate the nature of Marfa Lights phenomena, I needed to develop an investigative strategy. How does one go about investigating the unknown? Based on available literature (see introduction to Stories, Appendix A, for a list of published sources), I concluded that a number of smart people have tried investigating Marfa Lights without much success (unless you want to count the efforts carried out by folks wanting to prove that car lights on Highway 67 are mistakenly assumed to be mysterious). My childhood plan of triangulation to obtain a geographical location that could be visited in daylight was an obvious strategy but one that I felt sure had been tried multiple times by other people, apparently without tangible results. Spouse Marlene (who died in 2001, soon after this quest began) clarified this point nicely when she commented, "If triangulation can solve this puzzle it would already be solved." Hard to argue with that logic. I needed a more comprehensive approach than just triangulation.

Of course, determination of ML points of origin would still be important for multiple reasons. For example, if MLs are dependent on gases venting out of the ground, then it is reasonable to expect that points of origin would repeat. Repeating locations, once established, could be monitored and investigated for vapors, mineral content, and other associated and potentially causative factors. Even knowing which direction to look for MLs would be helpful. So the first objective in my strategy list would be to search for **ML Locations.**

Another objective would be to determine **ML Composition**. What is the chemical and/or electrical make-up of MLs? If MLs are physical (i.e., not mirages) then gaining insight into their chemical and/or electrical composition could lead to an understanding of what they are and what accounts for their existence.

It would also be important to study and record **ML Behaviors**. For example, in the November 2000 event (Story 10), the lights were stationary, they turned on and off, varied in intensity and sometimes changed colors. They may also have executed small vertical movements. These were all ML behaviors. Recording and cataloging such behaviors would be important to gaining insight into their true nature and possible origins. Gaining an understanding of ML characteristics would also be important to ML recognition and the ability to distinguish between ML and artificial light sources was critical to issues of data purity.

The fourth investigative objective would be **ML Related Events**. For example: My November 2000 sighting occurred on a dark-cold-windy-night. Would any of those observed conditions prove to be required for ML appearances? What about rain, moon position, solar wind speed, and Northern Lights? Study and recognition of ML Related Events might provide valuable clues to the true nature of MLs in addition to their potential for predicting ML appearances.

My strategy, then, could be summarized as four investigative objectives: Location, Composition, Behavior and Related Events. Each of these objectives would translate into equipment and trips into the field (i.e., time and money).

Preparations (Open Wallet)

Location – The single most important tool for determining location is a compass for obtaining magnetic bearings. Not just any compass will do because it must be sufficiently convenient to read accurately in the dark. Obtaining magnetic bearings sounds like a minor consideration, but accuracy of bearing measurements are critical to computed location errors. I had taken a compass to our Thanksgiving encounter. To take a bearing, I had to elevate the instrument to eye level and sight across it at the MLs. To read the instrument it was necessary to lower it (while trying to maintain the selected bearing) sufficiently to shine a light and read the compass. Obviously, this was an error inducing procedure. My first solution was to glue the compass to a tripod-mounted spotting scope. This solution worked reasonably well but required using a flash light to read the compass and that was a distraction. Also accuracy of the compass was less than desired.

A better solution came in the form of an electronic compass equipped with sights. With this instrument it was possible to sight across the compass at eye level, push a button to hold the selected bearing and then lower the compass for reading. The display was

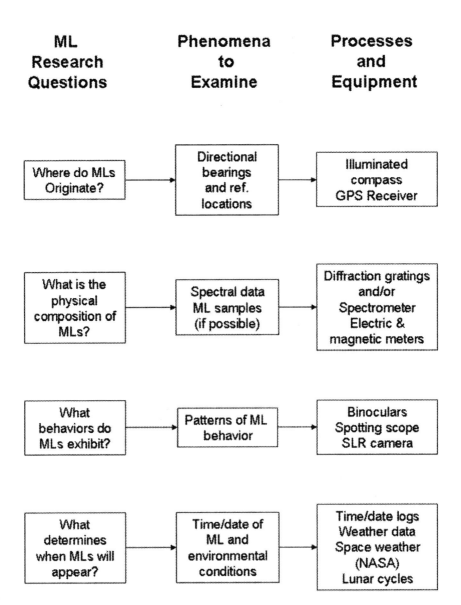

ML Research Questions	Phenomena to Examine	Processes and Equipment
Where do MLs Originate?	Directional bearings and ref. locations	Illuminated compass GPS Receiver
What is the physical composition of MLs?	Spectral data ML samples (if possible)	Diffraction gratings and/or Spectrometer Electric & magnetic meters
What behaviors do MLs exhibit?	Patterns of ML behavior	Binoculars Spotting scope SLR camera
What determines when MLs will appear?	Time/date of ML and environmental conditions	Time/date logs Weather data Space weather (NASA) Lunar cycles

Figure 17. Processes & equipment needed to investigate ML phenomena.

47

illuminated, perfect for night use, and the instrument was accurate and easy to read. Good as this solution was, it would eventually be replaced by Nikon binoculars with a built-in illuminated compass (Figure C10). This final solution was the best possible.

With an accurate method for taking bearings, range can be found by taking measurements from different locations and computing cross-bearing intersections (triangulation). It would also be possible to estimate ML altitudes by reference to background mesas that rise to about 500 feet above the valley floor.

Composition – How could I determine anything about ML composition without actually capturing one? One way would be to examine the ML's spectra (i.e., light frequencies). Unfortunately, spectrometers at that time were poorly suited to the task as well as beyond my limited investigative budget. However, my search for a suitable tool did turn up a small, inexpensive educational spectral device available from a science supply house. I ordered the device which was nothing more than a plastic capture box for light that entered through a diffraction grating filter, and tried it out in my house by "reading" fluorescent and incandescent light bulbs. The filter was scored to create thousands of tiny prisms that collectively parsed the light by frequency. The box also contained a scale for reading frequencies. The instrument was somewhat crude but ingenious and could distinguish between plasma and non-plasma lights. I could certainly see value in that because car lights use solid light filaments that produce continuous spectra. It was my hope that this small blue box would not only help me distinguish between car lights and MLs but, also provide clues about ML composition (light frequency information).

Behavior – The ML encounter in November 2000 gave me an initial set of behavioral information based on binocular-aided

48

observations. Good binoculars are essential tools because of their ability to bring distant objects close and their inherent ease of use. I started with 7 X 50 (i.e. seven power with 50 mm objective lens) that worked well in the dark but soon moved up to 10 X 50 binoculars (ten power being a practical limit for hand-held binoculars). Binoculars were supplemented with a 12X spotting scope that was only a little more powerful but probably twice as effective for evaluating distant lights because it was mounted on a stable tripod instead of hand-held.

Optical aids are great, but I also needed to photograph future ML encounters in order to create permanent records with more detail and accuracy. I already owned a perfectly good Sears film camera and it was still like new even if it was 41 years old. I had never gotten into photography and had no experience to draw on. Spouse Marlene urged me to purchase a modern Single Lens Reflex (SLR) camera. I knew she was right but was reluctant to spend that much money for what I expected to be at most a two to three month investigation. She convinced me that it was necessary because I would be photographing in the dark and my existing camera did not provide sufficient manual control for low light conditions. We went together to a camera store and I purchased a Pentax ZX-30 film model (this was before the age of digital SLR cameras). With the camera issue resolved, I made plans to return to Marfa in December 2000.

Looking back on that SLR purchase decision, I can only smile. It reminds me of the monk who one day decided it would do him good to get out and see the world. In preparation he loaded provisions on a donkey, opened the gates early one morning and set off to see the world. He found the unfolding countryside beautiful but riding the donkey was a little rough. He rode until he was tired and then got off and walked until he was tired of walking. Getting back on the donkey he continued the journey until he was ready to walk again. By late afternoon, he was one tired monk, but he was overjoyed at seeing what a magnificent world God had created. He stopped by a house to wipe his brow and ask for water. The owner of the house, seeing that the monk looked a little disheveled, inquired if everything was okay. The monk explained that he had set out to see the world but the world had turned out to be bigger than he thought it would be.

At the beginning of my ML investigation I was about as ignorant of what lay ahead as that monk. More time and treasure has been consumed since that first SLR purchase but, like the monk, I find the journey absolutely fascinating.

Related Events: More things to consider – It was a cold, windy night with no moon when I sighted the November 2000 MLs. It would be important to see which, if any, of those conditions would be repeated with future encounters and to look for other circumstances that might be related to ML appearances. One of my first thoughts regarding "connections" was the possibility that MLs might be a middle latitude version of the Northern Lights. I checked to see if the Northern Lights were active during the November 2000 ML events and discovered a high level of solar activity during that very time frame. Interesting! Fortunately, that was a pattern that would be relatively easy to track thanks to publication of NASA data on the internet. Looking into NASA data, I was delighted to find indices to help predict Northern Lights and many other parameters relating to solar weather. I would start referring to this data before leaving for Marfa and would keep up with it while there by visiting public libraries in Alpine or Marfa.

Maybe solar events would correlate with ML appearances in

Mitchell Flat and this would turn out to be a short investigation. But weather factors such as temperature, humidity, wind speed, precipitation, and barometric pressure were all possible contributors. I decided to start monitoring those conditions as reported by the weather service for the time of each observed ML.

I had a plan for investigation and had acquired a reasonable set of investigative tools. The next step would be to return to the field and start collecting data. I could hardly wait!

"Filmed them on my Sony camcorder. Highly doubt they were mirages. Behavior was best described as "undulating plasma balls" that folded in on each other, then split, blossomed, divided, rejoined, re-split, then dimmed, bloomed again, then blinked off. Like spherical torches being suddenly snuffed out. "-- from Story 7

Ventures Into Darkness

Trapped in the Night

It was December 2000 when I returned to Marfa and began my pursuit of MLs. There I was, fully equipped with spotting scope, compass, camera, gun sight (no gun, but the sight had cross hairs useful for detecting small horizontal and vertical movements), and binoculars. I may have been ready, but the MLs were not. Although I faithfully observed every night with my camera and scopes at the ready, nothing unusual showed. There were many car lights on Highway 67 and occasional ranch trucks in Mitchell Flat. Nothing appeared that I could call an ML. Being green to the search, I was surprised and disappointed. In November there had been plenty of ML activity on both nights. Now it was December and the show seemed to be over.

On the last night of my planned stay it was growing increasingly obvious that I would be returning home empty-handed. Damn! I decided it was time to try my luck on a county ranch road located west of the View Park.

I drove boldly across the first cattle guard and down Nopal ranch road in the dark of night. They call the region Mitchell Flat but, in truth, it is anything but flat. The terrain consists of gently rolling hills and, for the first five or six miles, Nopal Road goes up and down. I found the vistas to be more restrictive than those available at the View Park but wanted to give this more southerly location a fair chance.

By 2 AM on December 24th I was ready to give up the hunt. I had planned to drive home that day and needed sleep before attempting the 500 mile drive. I decided to return to the View Park for one final look and then head on to my motel in Alpine. Driving north toward US 90 I could see beams of light probing the sky somewhere ahead of me. Someone was headed my way. I assumed it was a late returning rancher but felt a little uneasy given the remoteness of my location and lateness of the hour. As I topped a small rise, the oncoming vehicle was directly ahead of me and approaching rapidly. I tried to spot official markings on the side of the vehicle as it shot by, but the glimpse was too fleeting. I watched in my rear view mirror and could see an SUV brake sharply, spin around and come charging back in my direction. Whoever was in this vehicle was clearly after me! By then I was within half a mile of US 90 and possible escape, but I could see two more vehicles parked by the cattle guard ahead, ready to block my way should I try to run for it.

Not good! I was very much hoping that these folks were border patrol agents. The vehicle following close behind turned on red lights so I pulled over and stopped. A Border Patrol (BP) officer approached out of the dark. Shining a light in my direction the officer, a woman, asked if I knew that I was on private property. She informed me that she was a United States Immigration Officer and wanted to know what I was doing there. I explained that I was doing research on Marfa Lights and had been careful to not stray from the county maintained roadway. She was unimpressed and stated they could see me from the Border Patrol Check Station on Highway 67. Her statement was interesting because I would later determine that was a physical impossibility. Had I tripped an unseen roadside vehicle detection sensor? The BP keeps a close watch on late night activity on all ranch roads, but it was also entirely possible that a rancher in Mitchell Flat had seen my car lights and called them. Once she was convinced that I was not a drug smuggler or operating in any illegal way, I was permitted to continue my journey.

The Border Patrol works closely with local ranchers and they support each other. Drug cartels in Mexico are a rapidly growing threat, not only in Mexico, but increasingly in this country as well. Were it not for our great Border Patrol force, towns near the border like Presidio, Marfa, and Alpine would soon be too dangerous to visit.

This discussion is a side track to my story of hunting light, but I realize that some of my readers may have their own encounters with the Border Patrol if they go out looking for Marfa Lights. Over the years I have been stopped several times by the Border Patrol and am always appreciative of professionalism in the conduct of their difficult and dangerous duties. So if you do get stopped, do not be concerned; these people are friends, not enemies.

Seeking Elusive Lights

I tried looking for MLs again in mid-January 2001, but saw nothing. I was beginning to realize that this quest might not be as easy as I had assumed it would be. I needed to be on station in Marfa for a solid week or, better yet, two weeks. Surely two weeks would be long enough to ensure an encounter. During the day I could bone up on previous attempts and information by visiting Sul Ross University in Alpine. I was aware from ML literature that Sul Ross students and teachers had mounted numerous expeditions to Mitchell Flat in attempts to solve the puzzle of Marfa Lights. I was betting that their library would contain a gold mine of data. Why reinvent the wheel? Taking advantage of previous work would help keep me out of blind alleys and serve as food for new ideas.

This next trip would be a more significant effort. Not only had I prepared to stay longer, but I had also expanded my bag of investigative tools. My location equipment would be augmented with a scientific calculator preprogrammed to do triangulation calculations, a more complete set of topographical charts, and a GPS receiver that would make it possible to record latitude/longitude coordinates for points of interest. The behavioral tool set would also be augmented with a Generation 1 Night Vision device. Night vision devices amplify light dramatically and display results on built-in cathode ray tubes. Because of light amplification, this night vision device had potential to detect very dim lights that might otherwise go unnoticed.

I was ready for the lights, but the question was, would MLs

show? Unfortunately they did not. Waiting for mystery lights to appear is like being on a police stakeout: lots of waiting and very little action, punctuated by short periods of very high awareness. Boredom was partially dispelled by observation of other light seekers at the View Park. Even in cold weather there would typically be eight to twelve visitors earlier in the evening. In most cases, these visitors would leave believing they had seen the fabled Marfa Lights when what they had actually seen were vehicle lights on Highway 67. I found talking to visitors entertaining. They mostly came from Texas, as you might expect, but there were plenty of visitors from all over the United States and, for that matter, from all over the world.

Automobile lights on Highway 67 were a distraction, so I stopped looking in that direction (southwest) altogether and focused my attention to the south and southeast. A majority of visitors, however, stayed locked on Highway 67 headlights. I remember one evening when a lady sitting in her RV stuck her head out of the window and started shouting at me to look to the right. "Over there! Look to your right! The lights are to your right!" I tried to ignore her directions but she continued, so I glanced back to see her with head and shoulders out of the RV and hanging onto the side mirror as she continued to shout with increasing volume. I smiled and did a little return wave to acknowledge her advisory, then continued scanning to the southeast. Following that she gave up, but I could hear her say to her husband, "Some people are too dumb to even be out here."

Every night I watched and waited with no sign of MLs. Some nights were moonlit, some were without moonlight, some were cloudy, some were clear, and most were windy. If moonlight or weather were factors, it was certainly not obvious.

Checking on my idea that MLs might be tied somehow to solar activity and the northern lights, I went to the Public Library in Alpine during the day and monitored NASA solar events data on the internet. As luck would have it, solar conditions were once again very active during my extended trip in Marfa, but that fact made no difference. I would visit the library and note that a solar wind shock wave was about to strike the magnetosphere, or else one already had, and would think, "Tonight will be the night they appear," but it did not happen.

I was disappointed to find that the Sul Ross Library had no scientific studies of the lights; there were no theses or student studies on the Marfa Lights. Sul Ross did have a compilation of local press clippings, although Marfa and Alpine public libraries both had larger collections, with Marfa having the largest. But these were personal accounts. No one had done a long-term, systematic study of mystery lights. So much for the research of others; I was on my own!

That long field trip, and others to follow, revealed to me the first important fact about Marfa Lights: they are uncommon events. This inconvenient fact is a major impediment to discovery. I would soon learn that MLs are complex phenomena capable of manifesting in many different forms. But scarcity of ML appearances severely limits investigative efforts by slowing data collection to a trickle. Statistical analysis is hampered by painfully small data sets. Accumulation of even minimal data seems to take forever. My nightly log was becoming a compilation of non-events unworthy of being written down. With nothing happening, observational nights became long indeed.

I would come to appreciate how critically important my November 2000 sighting (Story 10) had been. It not only launched me into this strange quest but, more important than that, it became the steel anchor sustaining my research effort. Without that spectacular 2000 sighting there would have been no books and no research effort. If the sighting had been more ordinary or explainable my research effort (if it happened at all) would have become only additional certification that Marfa Lights were the products of automobile makers coupled with over-imagination of tourists and Marfans lacking critical thought.

But the incredible ML displays Marlene and I had seen in November of 2000 did happen, they were real, and I was absolutely certain (still am) that they were not car lights or ranch lights, or stellar mirages, or phosphor coated rabbits, or teenagers waving flashlights, or any of a dozen other offered explanations. What they were was something truly unusual, not paranormal, something physical, but clearly one step beyond the world's inventory of known phenomena. Understanding that startling fact and recognizing the significance of ML existence is what has made it possible to wade through seemingly endless nights of boredom to seek the thrill of one more spectacular ML event. Additional events would come but always with painful slowness.

Did You See What I Imagined That I Saw?

After a time, I began to notice something that had at first missed my attention. Peering into the gathering darkness--and hoping to see something--can play tricks on the eye and the mind. In the twilight following sunset, before real darkness, I was seeing small fleeting specks of light. They were so small and brief that it was easy to believe they were tricks of eyesight. Did I really see that or only imagine that I saw it? It was easy to write off these events until I noticed that other observers were seeing them too. I would think that I saw a tiny speck of light shooting right to left and then hear someone else say, "I just saw a tiny light flying left!" If we both saw that speck of light, then it must have really happened. I started listening to what others were saying in this regard and it became apparent that these specks of light were not imaginary; they were real. They reminded me of sparks from a fire, but they seemed to be flying against the wind.

The fact that multiple people were seeing these fleeting light specks might have a logical explanation. They might be light reflections caused by passing traffic on US 90 lighting up bottle caps or other reflective materials. I looked for that possibility but concluded that these specks were in fact real, as tiny and short lived as they might be. What were they? I could find no way to photograph them because of their tiny size and short lifetimes. They would disappear before I could track and focus my SLR camera on them. They would burn only one time and flew consistently into the wind. The flight paths were horizontal to the ground; they did not go off in all directions. Were these light specks another manifestation of the larger phenomena known as Marfa Lights? It seemed likely that they were. I would later decide that MLs came in multiple forms and these tiny lights were one such form.

Night Photography Is an Art

Dogged persistence does pay off; after many nights of disappointment I did get to see an ML again. By then I was so jaded at seeing only ranch trucks in Mitchell Flat that I was slow to comprehend that the light was different. I was like a guy sitting half-asleep at a long stop light who is shocked by the realization that the light has finally turned green. I was jolted out of this numb state when the ML split into two parts with the weaker left component

60

winking on and off. I immediately leaped into action and ran to turn on my camera.

It was great to observe this miracle in action but what I wanted most was to record this behavior on camera. Photographing MLs does require more than pointing a still or video camera and pressing the shutter button. The internet is filled with photographs and YouTube videos taken by people who believe otherwise. From time to time I peruse the latest internet collection and rarely ever see anything other than pictures of car lights, usually moving in dark backgrounds devoid of terrain features.

Simply put, MLs are nocturnal events and photographing them requires light control. Consumer-level digital point-and-shoot cameras and video cameras do a remarkably good job taking pictures in daylight. At night the same devices are often severely light deprived. Getting as much light as possible into the camera is necessary to achieve meaningful ML captures. Cameras equipped with user selectable Shutter-Priority (Tv) and Aperture-Priority (Av) modes are essential. Single Lens Reflex (SLR) cameras are the preferred choice, but any camera with Tv and Av modes can probably be made to work.

The key to successful ML pictures is use of long exposure times in order to accumulate sufficient light to achieve good photographs in the dark of night. Long exposures open the door to camera artifacts making solid tripods and wire, cable, or remote shutter release methods indispensable. Even with a solid tripod and wired shutter release, it is still easy to pick up camera artifacts when targeted MLs are on the move because the camera operator must keep repositioning the camera. The right procedure would be to wait for motion induced oscillations to dampen out following camera movement, but in practice I could not wait for that dampening; ML opportunities are so rare and often very quick events. The price paid for maximizing possible picture opportunities was that it then became necessary to weed out vibration induced artifacts later.

Picture Disaster

These camera techniques seem simple enough in hindsight but getting that first ML photograph proved more of a challenge than I had anticipated. Sometimes we prepare for the big and then stumble on the small. My first attempt at capturing real MLs failed simply because I had not fully digested nuances of my SLR camera/lens combination.

When a second opportunity finally came I was able to make use of a new 300 mm telephoto lens purchased right before the trip. To ensure maximum light, I switched the lens from the "A" (for automatic mode) and manually rotated the lens to the full open position. Using a tripod and cable shutter release to avoid camera shake, I carefully aimed the camera at the ML and snapped away -- click, click, click, while the ML performed fascinating antics of splitting and merging. This was exiting! I could hardly wait to see what my new lens had captured. But first there was the drive home and then the wait for film processing. Finally I got to see the results of my efforts and learned that I had a collection of black photographs. Nothing came out.

What? Nothing? How can that be? Post-disaster analysis revealed that I should have read the owner's manual for the new lens more carefully. The moment I moved the aperture ring away from the "A" position the lens/camera system stopped working and took only black pictures that night. It was possible to operate the new lens in manual mode, but only by using the camera's LCD display to set aperture and shutter speed.

I felt dumb to have missed that vital instruction but in truth it was a mistake even a camera-savvy person might have made. The failure was disappointing, mostly because it had taken so long to find another ML event. When the next ML opportunity presented itself, I was completely ready and took a number of time exposures. I was sure the pictures would be good and could hardly wait for them to be developed.

The next morning, as I loaded my car, the camera back accidentally popped open. Quicker than Superman in a hurry, I closed it tight. Unfortunately, even Superman is slower than the speed of light. My hard-won photographs were ruined and the lengthy effort required to get them was wasted.

I kept returning to Marfa to look for MLs, and eventually another opportunity came my way on a night that was cold and made colder by a strong wind that chilled me to the bone. MLs pulsing across the prairie didn't seem to mind the wind and were putting on a great display while I collected time exposures, click, click. Then a gust of wind blew over my tripod. So what! I picked it up, brushed off the dirt with my hand, and continued clicking pictures in the dark. It was the next day in daylight that I learned the bad news. The fall had broken my lens and the morning light was already wandering around inside my camera. None of the pictures came out!

I spent my working career dealing with aerospace defense contracts. Hidden, top secret programs were known in the industry as "black programs" because they were like black holes; information could not escape and no one was supposed to know the programs even existed. I joked that my little hobby-project had become a black program too, but not in the usual sense. The black part had to do with ML photographs. There I was proclaiming the existence of exciting mysterious lights without a shred of photographic evidence to back that claim!

I have never been one to give up too easily. I purchased a five-pound weight to hang from my tripod to eliminate wind driven blow-over. To double my chances against another failure I purchased another Pentax and began using two cameras and two weighted tripods. Sitting through bitter-cold nights I pressed on, determined to get that photograph. Eventually another opportunity did come my way and I took a total of six photographs. This time my jinx slept and the photographs developed nicely. A couple of them were a little overexposed, but at least I had something. After that painfully slow start, photographing MLs was never again as difficult.

Because MLs are usually miles away from the View Park, I found my new 300 mm lens helpful because of its ability to magnify images six times. The advantages of 6X magnification were important to my research so I purchased an additional 300 mm lens for my second camera.

Some wise person once observed that "better" is the enemy of "good enough." If six power is good then ten power ought to be better. And, for that matter, why stop at ten power? Wanting ever more detailed pictures drove me to seek more powerful camera lenses. Soon telescopes would become my lens of choice and, over time, those telescopes would grow in size because my pursuit of more detailed information was relentless. Film cameras would give way to the marvel of modern digital cameras and I would even modify one of those to give it the ability to record infrared (IR) frequencies in addition to visual frequencies.

A Personal Note

Wife Marlene had urged me to buy that first SLR camera and I suspect she was smiling from heaven as she watched my ever expanding collection of investigative equipment. Sadly, Marlene died in 2001 before this project began to take shape.

Several years later my hunt for mystery lights was taking me to Marfa frequently, making it convenient for me to attend my June 2004 Marfa High School reunion. I don't know if it was planned in Heaven or simply excellent fortune, but attending that reunion was the best decision of my life. It was there that I met my current wife, Sandra Martin Dees (Sandy). We remembered each other from our high school days. We had both lost our lifetime partners. She bought my book, *Night Orbs*, and then emailed me with questions about my data (she is a research psychologist). Our conversations about the Marfa Lights project and about the partners we had lost were a gift to us both. We married in March of 2007.

We Needed to Solve the Puzzle In Order to Solve the Puzzle

Nights in Marfa were devoted to the quest for more sightings while daytime activities were split between library and internet searches plus discussions with those who had an interest in the lights, often with stories of their own to tell. The stories presented in Appendix A are a representative sampling, but there are countless similar stories. By January 2002, I was starting to appreciate the wide scope and extreme depth of this intriguing mystery. There was something fundamentally tangible and real at work; getting to the bottom of it would not be easy.

The night vision device I had purchased proved to be very light sensitive, making it possible to find minor levels of light, but good binoculars and other optical aids were able to do the same and were generally easier to use. The spectrometer tool I was carrying with me had been designed for educational laboratory work in a lighted room. I had no trouble using it in daylight but, in the dark of night, the instrument would be useless unless the emitting light was bright enough to illuminate the internal scale. I carried this instrument with me for a few trips to Marfa but soon faced the reality that it was unlikely to work unless an ML were to appear within the confines of the View Park, so I stopped carrying it. Giving up on this limited spectral tool was not the end of my attempts to seek ML composition as will be discussed later.

The biggest obstacle to data collection continued to be the relative rarity of ML events. I needed to predict when MLs would appear in order to collect enough data to identify ML related events, but I needed to find ML related events in order to predict when MLs would appear. It was a classic Catch 22. I needed to solve the puzzle in order to solve the puzzle. The only alternative was painfully slow data acquisition based on nothing more than luck to achieve ML encounters and they were, I was finding, few and far between.

Fortunately, trips to investigate the lights were fun, but in 2002 I estimated the chances of seeing MLs on any given night were probably no better than one in ten. Today I would estimate even less favorable odds. A college professor and friend of mine, Dr. Karl Stephan, spent 20 continuous nights at the View Park starting in May 2008 without seeing a single ML. If you are thinking of making a trip to Marfa to see firsthand what this is all about, do not let my gloomy assessment of poor odds slow you down. It will be a fun and enjoyable trip -- maybe even exciting -- no matter what the outcome.

MLs Race Cross-country

My initial expectations about MLs may have been limited to some extent by stationary aspects of the ML sightings experienced with Marlene in November 2000. In November 2002, I witnessed three MLs on two nights, including one that moved cross-country for miles. When I noticed the first ML it was rapidly moving west, across Mitchell Flat, I had assumed it to be a ranch truck on Nopal Road. However, the light appeared to be pulsing or flickering and that captured my attention. I wondered if this might be some kind of atmospheric flicker or maybe truck lights obstructed by brush, but I could not see any light beam from headlights or the flash of taillights that typically display when trucks must slow to go over cattle guards.

Behavior -- As the ML advanced closer to my location at the View Park it became obvious that it was following a path where no road exists at a speed not possible without a road. Also, this approaching light turned off completely two or three times, and remained off for a few seconds before shining again. Each time, reappearance would begin with a brilliant flash of light as if the ML was somehow recharged while in an off state.

How was this ML able to go out completely for seconds and then, with a bright flash of light, resume its journey in the same direction as before, and then repeat this cycle multiple times while traveling for miles? That was hard to understand and it raised an important question: Where did ML energy come from, especially when MLs are on the move and traveling for miles?

There were two possibilities:

1. MLs "run" on internal energy that lasts only as long as they do, or

2. MLs are maintained by an external source of energy.

67

The latter concept seemed to best fit what I had observed. As the ML dashed cross-country, it would slow, dim and then go out as if it was losing energy. After a few seconds or minutes, it would reappear with a bright flash and resume motion, as if its power source had been replenished. If MLs have all of their energy at birth, then what would account for their cyclic decay to off states, and then back on to shine brightly? Also, it seemed so unlikely that MLs would have enough internal energy at birth to function so dynamically for significant periods of time and to travel for miles cross-country with repeating energy cycles.

But if MLs were being fed additional energy as they traveled, where in the world could that energy be coming from? The ML puzzle was getting deeper. These 2002 MLs forced significant expansion of observed ML characteristics. My uncle John was right when he observed that MLs are not always stationary; they can move cross-country for long distances. They vary in intensity and turn on and off even while moving. Off states are usually followed by bright flashes as they resume their journey. Points of origin are varied. Directions of travel appear to be independent from wind direction. It was becoming obvious that MLs are complex phenomena, but I still had much to learn. With continued study, ML complexities would continue to expand driving the mysterious nature of these phenomena ever deeper.

"After observing for 10 or 15 minutes they noticed a light about 15 degrees above the skyline and a little left of center be- tween two Mercury Vapor (MV) lights. It would climb in a scooping motion, and then come down as if it were rolling down stairs. It moved very suddenly from left to right and then back again, all the while becoming brighter and dimmer and changing colors from yellow-white to red." -- from Story 12

Monitoring Stations

Roofus -- January 2003

Trips to Marfa were expensive, time consuming and, in most cases, unproductive because MLs were so rare and I had found no patterns to predict when they would occur. Even if there were detectable patterns to be found, I worried that with my rate of discovery it might take two lifetimes to get enough data to find those patterns. My research needed onsite support to build a log of ML events. I needed cameras that stayed in Mitchell Flat.

I am fortunate to still have several friends from my early days in Marfa. One of them owns a ranch with a great view of Mitchell

Flat from their ranch headquarters. I asked his permission to mount a security camera on a pole that I would plant in some mutually agreed to location. The ranch owner had a better idea and suggested that I mount it on the roof of one of their buildings. Knowing how strong the wind blows, he said it would be okay to nail it to the roof. Wow! The roof idea was much better and probably necessary because it would provide enough elevation to see over a nearby hill.

But I could not see putting nails in the roof. Having grown up in West Texas, I was well aware that he was right about high winds that sometimes blow at near hurricane velocities. The challenge was to create a platform, without nails, one that would stay put in those strong winds. I developed a saddle design that would fit over the roof apex and be held in place with weights, using no nails or screws. Installed in January 2003, the design worked, the camera was stable, and is still stable and working (Figure 18).

As had been the case with my tripod-mounted cameras, this monitoring station came with a learning curve. I was used to buying telescopes, binoculars and consumer cameras that had coated lens. Most security cameras, I would learn, dispense with such amenities. Without coated lenses, all of my early video light targets had a windblown look caused by lens flare (lens flares are created when light enters the lens at such an angle that the light rays do not completely flow through the lens, but instead are reflected back and forth between lens elements), and they always seemed to be out of focus. I could not understand why achieving good focus was so difficult. I would go up on the roof and carefully focus the camera on a distant mountain or other reasonable target. Real time images feeding to my monitor would seem to be sharp and well defined. I would think, "This time I got it right" only to

Figure 18. Roofus camera housing and roof mount.

find out-of-focus, windblown, and generally lousy results when I returned to retrieve data. I had no prior experience with night video and it took awhile for me to deduce that nighttime has a higher percentage of infrared (IR) frequencies requiring a different focus at night. Consumer cameras, even cheap ones, avoid this problem by installing an IR filter to block infrared frequencies. But security cameras do not anticipate being used in applications where that is a problem, so they have no IR filter and must be focused at night if they are going to be used at night (with the exception of a few security lenses specially designed for day/night use).

Once I understood that problem, I climbed up on the roof in the dark and focused the camera. Results were much improved but still windblown because of lens flares caused by uncoated lenses. Upgrading the system by installing a high resolution video camera equipped with a multicoated lens resolved this minor problem and resulted in better (but mostly dark) nighttime imaging. The new camera was used successfully for a couple of years until I upgraded again to an "astronomer's" camera with superior nighttime performance.

I replaced the time-lapse recorders with digital video recorders (DVRs). Two were required. One would be left in Marfa while I kept the other at home for data review. Data review proved to be challenging because I was taking home 80 gigabytes of images with each data retrieval (this amount of data seemed large at the time but total image data collection would later grow to well over a terabyte). The Sony DVR had a built-in capability to detect motion but, unfortunately, the motion detection grid was an 8 X 10 array of points. It was ample for detecting an intruder in your store or garage, but worthless for detecting points of light ten miles away. So I began the process of data retrieval and manual review. It was a boring, time-consuming activity, but still superior to making endless trips to Marfa.

The first couple of years I had to go up on that roof many times and experiment with different lens configurations but eventually the system evolved into something reliable, and pretty much trouble free. At first, this camera would be known as ML Monitoring Station 1 or MLMS1; I would later give it a more colorful name, Roofus. Roofus is a great research tool and it continues collecting nightly video of Mitchell Flat to this very day. Quality of the recorded images today is about a thousand times better than was achieved in the beginning. But even early on, MLMS1 (Roofus) provided a quantum leap forward for my investigative effort.

Snoopy -- 2004

Roofus (MLMS1), was numbered "1" for a reason. I very much wanted an MLMS2 because it not only takes two to tango, it also takes two to triangulate. I was doing a lot of guessing about where MLs were located. I could derive bearing information with one camera but had not a clue how far away those recorded lights might be. A second monitoring station would give me a cross bearing, making it possible to estimate range and location.

MLMS2 (now known as Snoopy) had a surprise beginning. It was a typical hot summer day and I stopped at the Marfa Book Company to cool off. A rancher on the stool next to me, wearing his Stetson hat, was sipping a large ice-filled coffee drink of some kind. It looked so delicious that I had to ask what he was having. Ensuing conversation led to a discussion of my passionate pursuit of Marfa Lights and he told me of his own ML sightings. They had been from his mountain ranch. "I have a good view of all of Mitchell Flat," he said. "Really?" I replied with growing interest. Naturally I asked him if he would consider hosting Snoopy and he agreed.

The initial location for Snoopy had power and a complete, but distant, view of Mitchell Flat. I tried to compensate for the long range view by using a magnified camera lens but results were unsatisfactory. I considered moving to a closer location, but getting power to the site was an issue. The ranch owner suggested solving the power problem by going solar. He was using solar panels to power electric fences and they worked well. It all made sense to me. I could mount a solar panel on the side of Snoopy. However, this new site would require clearing brush and constructing a welded wire enclosure to keep cattle from destroying the system by using the panels for back scratching. Marfa friend Kerr Mitchell saved the day by offering to help with this relocation and made it all possible by using his equipment to prepare the site and drill post holes. He trailered his equipment to the site, cleared a spot and even took his trailer to town to haul the welded wire. Without his magnificent help and support, Snoopy would not exist. Together we prepared the system.

I looked forward to cheap trouble-free operation, but developing a reliable power generation system proved to be more complex than initially anticipated. Over time, and with advise from my electrical engineer brother and other good minds, I worked to improve Snoopy's performance and reliabilitiy. Power generation was enhanced by doubling the number of solar panels and by installing a wind generator. The wind generator complimented Snoopy's solar arrays with its ability to run at night and on cloudy

Figure 19. Snoopy has four solar panels and a wind driven generator for power.

days. Deep cycle marine batteries would give way to sealed AGM batteries that were better suited to hot weather operations. Power control circuitry was made more sophisticated to enhance system reliability.

During the 2004 time frame, I showed Snoopy (I was still calling it MLMS2) to one of my California friends, Linda Lorenzetti. Linda took one look at MLMS2 and said, "You should call it Snoopy because it is built on a dog house." I loved her suggestion and MLMS2 became Snoopy. That also motivated me to rename MLMS1, Roofus.

To achieve a fully mature autonomous monitoring system required many months and numerous upgrades, but the Snoopy that finally emerged is reliable and productive. Snoopy was where I

Figure 20. Dollface and her pet Owloween do a good job of keeping birds from decorating the solar panels.

first experimented with an astronomy type camera capable of stacking images to accumulate light and allowing user control over light sensitivity and contrast. With the help of even a little moonlight, these remarkable cameras are capable of capturing background terrain and they are sensitive enough to detect even the faintest of lights. This was exactly what was needed.

Having two monitoring stations opened the door to triangulation of ML locations and paved the way for better understanding of ML start points. This also enabled ground searches at computed points of origin to see if anything unusual could be found. Roofus and Snoopy working together would soon determine that MLs had multiple points of origin. In fact, MLs rarely originated from the same spot. That fact was an argument against gas venting because we would expect ground vents to be in fixed locations.

Figure 21. Snoopy was given a second camera in August 2008.

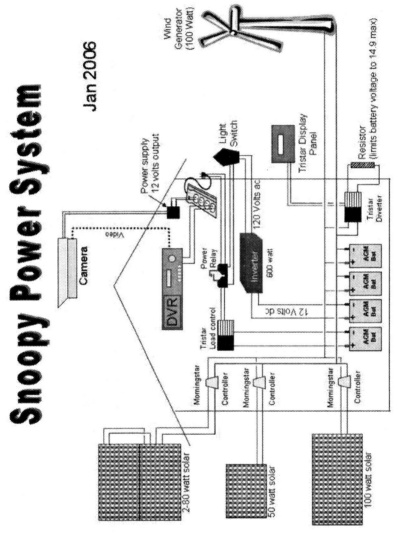

Figure 22. A graphic summary of Snoopy's electrical system.

77

Owlbert -- April 2006

Roofus and Snoopy working together had given me a way to calculate the locations of MLs but in order to do triangulation, the two camera fields of view had to have overlapping coverage. This requirement limited the extent of coverage possible in Mitchell Flat, but the scheme has worked successfully for a limited number of ML events. Computed positional errors were elliptical in shape with small uncertainty in the minor axis, and as much as 250 to 1000 feet uncertainty along the major axis, depending on range to the ML, precision of computed bearings, and relative look angles (see Figure 23). Not too bad, considering the distances and angles involved, but greater accuracy is always better. Navigators are taught to try and obtain three bearing fixes because, in most cases, having three bearing fixes reduces the error basket to a significant degree. Also, having a third station would provide for broader coverage than was possible with only two cameras. Knowing where I wanted to place a new monitoring station, I contacted the ranch owners. Fortunately, they were interested in my work, and agreed to allow two of us (Sandy and myself) to build a monitoring station on their property. We were delighted and very appreciative of the trust extended to us.

We located a building that was perfect for our purposes and received approval from the ranch owners. Cameras at this location would have the advantage of electrical power. But there was a strange, almost creepy, aspect to the building we had selected. There were many small bones, a few larger rabbit parts, and small animal skulls in and around the building. Death seemed to be everywhere. There were also strange egg-sized droppings from some animal I did not recognize.

Our second trip to the ranch revealed the source of all that

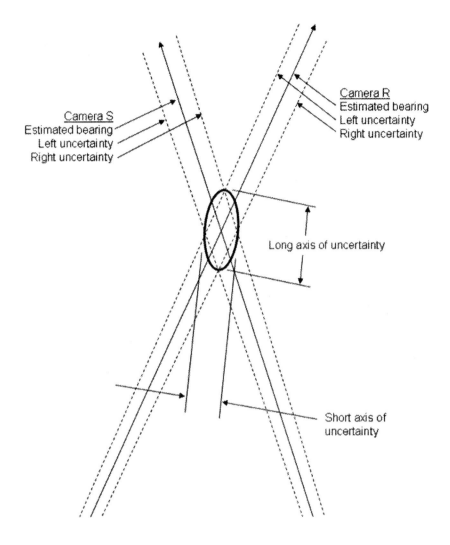

Figure 23. All measurements have error. The uncertainty ellipse for ML ground locations is a function of distance, accuracy of directional bearings and specific look angles.

strange stuff. The building was an owls' den! For some reason the owls were not in residence when we first visited, but on the second visit we would see at least four different species of owls living in the rafters of this building. The front opening was large enough for them to enter and they came and went as they pleased. The piles of bones and small skulls were remains of some of their meals. So were the strange egg-sized droppings. Those droppings came out of the front end, not the bottom! Owls deal with bones and other indigestible parts by coughing them up, like a cat coughs up hairballs. These owls are really magnificent creatures, and we enjoyed watching and photographing them a lot more than they enjoyed us. In fact, they must have hated our invasion of their previously peaceful domain.

We needed a suitable name for this new monitoring station and Sandy suggested naming it "Owlbert," given that it was to be located in an owls' den and the ranch manager was named Bert. That was perfect, and thus began our creation of the Owlbert monitoring station. I decided to go with multiple cameras equipped with variable magnification lenses to permit closer observation of any regions that proved to be ML active. Creation of Owlbert was a little less demanding than Snoopy had been, but not by much. Sandy and I accomplished the build in 16 twelve hour days, spread over four Marfa trips, to create a two-part station, Owlbert A and Owlbert B, located at different extremes of the host building.

I had tried using a color camera at Snoopy but had been disappointed by poor nighttime performance. At Owlbert I again tried using a mix of color cameras based on the manufacturer's lux specifications, but once again was disappointed with weak performance and reverted to all black & white cameras. Today Owlbert has five astronomy cameras plus one very low lux security camera that provides coverage for the first half hour after sunset and the last half hour before sunrise, when my more sensitive astronomy cameras are "whited-out" by ambient light. The primary cameras only become usable with increased darkness.

Figure 24. "What are these awful people doing in our house?"

Figure 25. Canine assistant

Figure 26. Owlbert A

Figure 27. Owlbert B

82

Figure 28. Owls believe in lofty living.

Figure 29. This owl tried to take my small dog to lunch. All the doors on my truck were open and this owl headed straight for my small dog in the back seat (see Fig. 25), but I scared him into an abort. He climbed sharply to clear the truck and then stalled. He managed to grab a beam as he fell where he hung by one claw looking back to see if I was after him. Before he got off the beam I closed the truck doors and then he moved to a beam directly in front of my vehicle where this picture was taken. You will notice his unblinking stare. He was looking for another opportunity but did not get one.

83

"After a short period of back and forth dancing, the right ML started moving rapidly to the west and continued its journey for miles, passing two mercury vapor (MV) lights en route. The left ML continued to oscillate and vary in brightness (pulsing) and exhibited interesting behavior including jets of material that moved down and to the right" -- from Story 17

ML Behaviors

Optically Close

On February 19, 2003, I had the good fortune of photographing an ML at close range even though I failed to realize that accomplishment at the time. My field notes from that night read as follows (these notes reflect measurements taken at the view park):

Visibility 9 miles, mostly cloudy

Weather predicts 50% chance of rain [it did not rain]

Baro 30.13, 44% humidity, ceiling 7000 feet, dew pt. 34 F

Sunset at 6:46 PM CST

55 deg F at 4:44 (temp. dropped sharply after sunset)

Wind NE at 14 mph, gusting to 22 mph
6:15 PM Wind NE at 18 mph with gusts to 24 mph.

This strong, cold wind may have decayed some by 8:20 PM but my notes did not record temperature or wind velocity at that time. They do note appearance of a light moving west from 174 degrees magnetic compass direction to 186 degrees. These field notes simply say:

Westbound light seen & photographed prior to 8:30 PM.

I took three photographs of that westbound light and was overjoyed when they were developed (see Figures 30-32). It turned out that this encounter was as optically close as I have ever been. To my naked eyes, this ML looked like an amber ball of light but my magnified camera images show evidence of what looks to be chemical combustion processes. These three time exposures are well worth studying because they display a running series of eruptions complete with off states and step changes in brightness so characteristic of MLs. If you could get close enough, all chemical-based MLs might show the same detail. Later in my investigation I would find other examples of possible flames and violent combustion processes, but these early film camera photographs are still the best and possibly the most significant ML photographs I have taken because they provide evidence relevant to ML composition.

There was now reason to suspect that MLs may be chemical combustion processes (i.e., chemical oxidation), although I was not yet ready, on the basis of these three photographs, to rule out the possibility that MLs might be a more energetic form of matter known as plasma. Plasma is completely ionized gas, composed

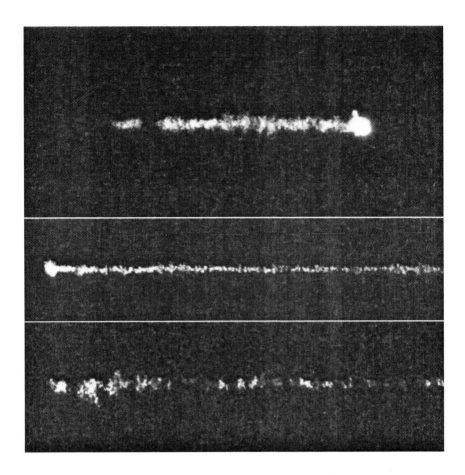

Figures 30, 31, & 32. This was a single ML photographed at ~8:20 PM CST on 19 Feb. 2003. At this writing, these time exposure film photographs are the most detailed ever taken. To my naked eyes, this was a ball of light traveling west with on and off states as well as occasional bursts to brightness. These three photographs reveal marvelous ML detail with the appearance of chemical combustion including at least two re-ignitions and step changes in brightness. This level of detail is inconsistent with what might be expected from motor vehicle lights or mirages of motor vehicle lights.

entirely of a nearly equal number of positive and negative free charges (positive ions and electrons). Plasma is the most energetic state of matter and the most common form of matter in the universe (stars are plasma). I was inclined to suspect that MLs might be plasma because their dynamic behavior and capacity for intense displays indicates that they are high-energy events. More research was needed!

April Showers Bring May...MLs?

Who knows if April showers had anything to do with it? What mattered to me was getting to witness and photograph extraordinary ML events on May 7 and 8, 2003 (Figures 33 to 36). They were the most spectacular ML sightings since November 2000. These two events are summarized as Stories 17 and 18, but following are additional relevant details:

May 7, 2003 – Sunset was at 8:36 PM, temperature was about 70 degrees F, the air was calm, the skies were clear with no precipitation and the ground was dry. About 30 minutes after sunset an ML appeared to the southeast and flared to high brightness. It was noticeably brighter than the brightest mercury vapor (MV) ranch light but before I could take a picture it dimmed and went out.

At about 9:55 PM CDT two new MLs (A and B) appeared with the left one in about the same location as the initial ML. The two were separated by ~1/4 degree with the right (B) light being the brightest. ML (B) pulsed brightly while ML (A) remained steady. Both of my cameras were taking time exposures. At 9:59 PM CDT ML (B) began moving to the right (west) and continued to do so right on past both the center mercury vapor (MV) and the west MV ranch lights. After passing the west MV, another ML (C) appeared and it followed ML (B). Meanwhile ML (A) was holding position, pulsing, and had become brighter. ML (B), with ML (C) following at a distance, traveled another 6 to 8 angular degrees west of the west MV light and then reversed course along with ML (C). Both MLs (B) and (C) moved back near the MV ranch lights

Figure 33. MLs on 7 May 2003 located southeast of the View Park near Whirlwind Mesa. This time exposure was taken at 9:56 PM but lingering twilight and computer enhancement have permitted sufficient light collection to illuminate terrain. The MLs were splitting and moving during this time exposure to create the light shape shown. They are estimated to have been flying at an altitude of 50 to 100 feet above terrain based on the height of background mesas. This was the beginning of an extensive ML light display as shown in the following photographs.

89

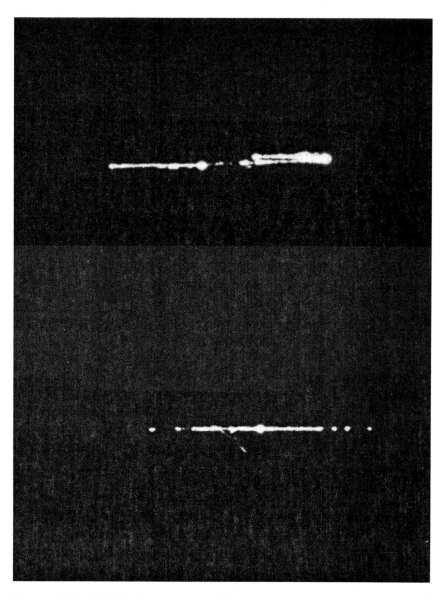

Figures 34 & 35. Splitting MLs moved right and danced back and forth as shown in the top figure. Another ML moved further right (westward) and generated offspring that orbited, or else oscillated back and forth, while the primary ML remained stationary at the center as shown in the lower figure. Notice the two jets of falling material. While these MLs danced, two other generated MLs continued traveling right for many miles.

90

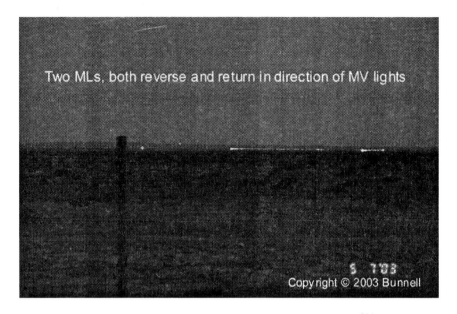

Figure 36. After traveling northwest, these two MLs reversed direction and traveled eastward but went out short of coming back in line with the two Mercury Vapor (MV) ranch lights.

and then went out at the same time. ML (A) still remained dancing in the southeast and was the last to extinguish. By 10:40 PM all MLs were out for the night.

I checked ranch lights, Marfa city lights, and airport lights to see if there were any hints of mirage conditions but there were none. This had been a spectacular display and I was delighted to have captured these amazing activities on film.

May 8, 2003 – As the sun set there was light wind with a few scattered clouds but skies were clearing. There were no indications of thunderstorms in any direction. Temperature was around 60 degrees F. At 10:16 PM a possible ML flared briefly in a location near that of the ML (A) on the previous night. At 10:22 PM the ML returned; I will call this ML (D).

ML (D) began moving west in the same fashion as ML (B) on

91

5/7/2003, but this time only a single ML was observed. Magnetic bearing at the start point was 153 degrees; at extreme finish the bearing was 208 degrees magnetic. This night the ML traveled further west but did not reverse direction. The final location was west of the railroad tracks, where it was observed by Rancher Kerr Mitchell. He noted that it was bigger and brighter than any of his previous ML sightings.

MLs on these two nights were also captured by Roofus. The fact that I managed to obtain start and end bearings from the View Park coupled with Roofus coverage made it possible to triangulate start and stop locations. The distance between the start and stop points measured eleven miles and that remains the longest computed ML travel distance. This ML event is also unique because of the explosive-like expansion that resulted in the light going out and then resuming at a much lower altitude (see Figures 37 & 38).

Location -- Origin of MLs on these two nights was southeast of the View Park near Whirlwind Mesa. This was the first time that I was able to triangulate start and stop locations providing terrain-specific location information. The path traveled between the start and stop point could not be computed but may have been a straight line.

Behavior -- These ML events expanded my list of observed behaviors with clear evidence of multiple splitting, merging, orbiting and/or dancing, evidence of descending material, long distance travel and the explosive expansion photographed on May 8th. Recorded ML behavior was becoming increasingly complex and ever more mysterious.

Related Events -- These May 2003 ML events happened in warmer weather as previously noted. Wind speed was 4.6 mph on May 7th and 10.4 mph on May 8th. Humidity was 11% on both

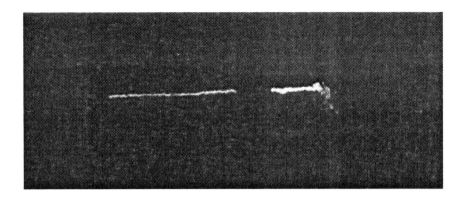

Figure 37. The very next night, 8 May 2003, MLs again put on a spectacular display moving left to right. This photograph is the best from that night. It shows the ML is flying at an altitude of about 250 feet based on background mesas. At that height the gaps would be caused by the ML going completely out and then back on. Notice the explosive ending. The ML went out at that point but came back on a few minutes later at a significantly lower altitude (see Figure 38).

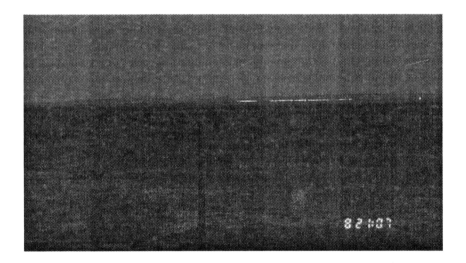

Figure 38. This less magnified view shows the 8 May 2008 ML continuing westward at a lower altitude. This ML traveled eleven miles to the Northwest but then went out without reversing course.

93

nights and there was no precipitation on either night. There were no Earth-directed solar Coronal Mass Ejections (CMEs) within nine days prior to either event.

May Surprise

By May, 2004, I was growing concerned about not seeing MLs for a prolonged period (there had been none since September 2003). Once again, the chance would come when I least expected it. On my last night of that May trip, I ended up at the View Park and experienced another outstanding ML display as summarized in Story 22, Appendix A. The previous September I had captured discontinuous ML spectra (as will be discussed in the next chapter). The finding of plasma-type spectra created an expectation that all MLs were plasma and that they could be easily distinguished from headlights by simply checking for discontinuous spectra.

But to my surprise, May 2004 MLs did not conform to this expectation. While giving a talk to college students at the View Park, I dismissed the first ML appearance believing it to be a truck light after seeing evidence of continuous spectra in my binoculars. However, the pulsing and flashes to exteme brightness soon convinced me that we were observing an ML. I rushed back to my truck, pulled out a camera and tripod too late to photograph the first ML but in time to take five photographs of two new MLs that followed the first one. Oddly, the first picture that I managed to take showed no spectra at all (Figure 39) even though it was the same camera and lens used to take the last three time exposures (Figures 40-42) and they all clearly showed continuous spectra (spectral patterns are visible left and right of the ML in Figure 40, right of the ML in Figure 41 and left of the ML in Figure 42).

There was no question in my mind that the lights observed

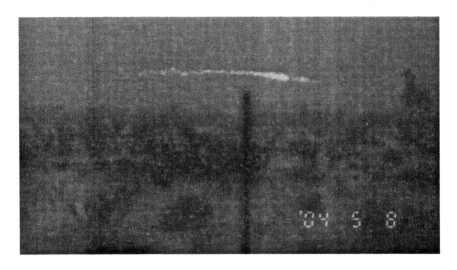

Figure 39. Time exposure of ML flying northwest into a strong wind on 8 May 2004. The second photograph in this series is not shown.

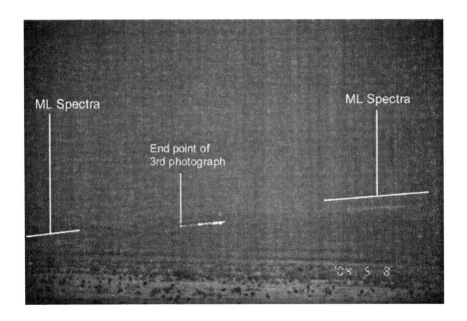

Figure 40. This third photograph shows the third ML moving down and to the left. Last point is flagged as a common reference point for next two photographs.

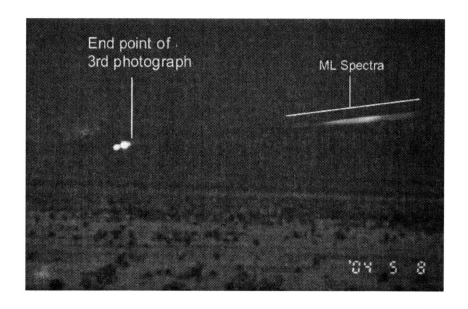

Figure 41. MLs slowed but continued moving down and to the left while pulsing on and off in typical ML fashion.

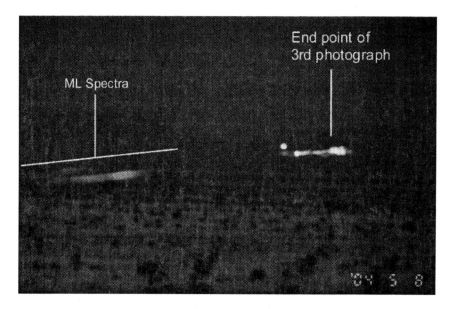

Figure 42. MLs moved back and forth as the reference line reveals. Final point was the highest light on the left side of the light track.

and photographed that night in May were MLs because of their pulsing behaviors and cross-country travel speeds in locations where there are no roads. Three of the photographs clearly showed continuous spectra, indicating they were not plasma lights. It was unthinkable that MLs could be appearing in both plasma and non-plasma forms and look much the same either way.

There were three possibilities: 1) The September 2003 light was a mobile manmade plasma light, 2) the lights I photographed in May 2004 were vehicles, or 3) all of these lights were MLs. I felt certain that the May 2004 event involved authentic MLs. I had longer to observe in May 2004 than had been the case in September 2003, there were multiple lights in 2004, and their behavior was consistent with previous ML encounters. The September sighting was rushed and I had time for only one hurriedly taken photograph. On the other hand, the September 2003 sighting was supported by measured electromagnetic anomalies and it had been traveling rapidly in a location where there were no roads. That left possibility 3, but if that were true, then it would mean that MLs could be either chemical combustion or plasma. There did not seem to be any clear way to know which possibility was correct.

The ML mystery was getting deeper. ML location, composition, and the range of behavior characteristics were expanding instead of narrowing. This was not what I expected.

Electric MLs -- Story 26
Date: June 2nd and 3rd, 2005
Source: Roofus and Snoopy
Witnesses: Roofus, Snoopy and unknown military pilots

In June 2005, Roofus and Snoopy again proved their worth by

accurately recording the most spectacular ML events in my files. ML activity began shortly before midnight on June 2nd (CDT) and continued until after 3:42 AM on June 3rd. The events of that night took place in the middle of a violent thunderstorm and faded as the storm moved off to the northeast. To my surprise, ML activity seemed to be initiated by and associated with that raging electrical storm.

The sequence of ML events began 21 minutes before midnight when a small ML appeared to the right of the mercury vapor (MV) ranch light. Forty-three seconds later, a second ML appeared to the left of the MV light. In less than four minutes the number of MLs would grow to five, with three left and two right of the MV light. Most of these MLs appeared in connection with, or at least simultaneously with, lightning flashes.

During these ML displays, a camera-visible fringe appeared and grew thick on the tree located near Roofus. One by one, the MLs started going out and were all out by 11:43 PM CDT but the night's events were only just beginning.

A new ML was added and it became more intense. As this ML was growing in brightness a halo began to form over it. The halo was an arch of light extending between locations of the first two MLs. At its peak, the Halo ML was about five times brighter than the MV light (see Figure 43). Both Roofus and Snoopy had clear views of these events, making it possible to calculate halo dimensions (Figure 44).

The halo ML made two appearances and then disappeared completely in connection with the 100th lightning flash at 11:59 PM CDT (the first flash was at 11:38:39). The halo started at 11:51 PM and lasted for a total of eight minutes.

At 12:32 AM CDT on June 3rd, an earlier ML returned and

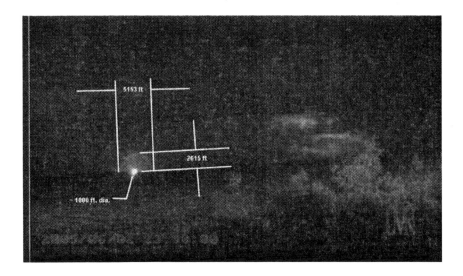

Figures 43 & 44. MLs on 2 June 2005 occurred during a raging thunderstorm and may be the most spectacular ever photographed. There are three MLs. The one in the center is dominate but the halo connects two flanking MLs. This complex ML display was 8.3 miles from Roofus and lasted 8 minutes. Height and width of the halo are reasonably accurate, but diameter of the central light ball is based on the bloomed image. The actual size of the center ball of light would be less.

Figure 45. This ML appeared 36 minutes later in about the same direction but 15 miles from Roofus. It had no halo but was bright enough to illuminate overhead clouds. The small light to the right is a ranch light located five miles closer to Roofus. This ML lasted >3 hours in the early hours of 3 June 2005.

Figure 46. One of three eastbound aircraft that flew over the ML of 3 June 2005. Pilots must have been impressed because three aircraft appeared from the east an hour later and dived down for a closer look.

100

seemed to grow bigger and brighter with each lightning strike. It would last for a total of three hours and six minutes and was bright enough to illuminate overhead clouds (Figure 45). Three eastbound aircraft flew over at fairly low altitudes during the time this big and very bright ML was shining (Figure 46). They were probably military because three aircraft (returning?) would arrive from the east an hour later and dive down to fly over the location where the final large ML had been located. It seems likely to me that three aircraft moving at jet speeds and diving close to the ground at night are more likely to be military than civilian.

These ML events were most unusual. To my knowledge, nothing similar has been reported by others nor have similar activities occurred since. These ML events are unique in a number of important ways:

1. They occurred during an electrical storm (rare).

2. They seemed to be fed by the storm (not seen before).

3. They were on continuously for extended periods without typical off states, but did vary in intensity (longest continuous "on" states that I have recorded).

4. They were all stationary MLs and did not convert to moving MLs (unusual for the number of MLs and time durations).

5. One ML developed a distinct halo that extended between two earlier MLs.

6. The Halo ML and the last ML were the two largest MLs I have recorded to date.

7. Size, brightness, and the 3+ hour duration of the last ML were unusual by any measure.

Triangulation permitted me to identify where these MLs had been located. Ground visitation of the Halo ML site revealed nothing that could explain the presence of these remarkable displays. This has also been the case of all ML sites that I have been able to visit. These electrical MLs provided valuable new information, but they also deepened this already deep mystery.

Infrared MLs -- August/October 2006

On August 11, 2006, I had another good sighting (Story 27) and was able to take a number of photographs with an infrared (IR) capable camera (Figures 47 and 48 are examples).

Two months later, on the night of October 19, 2006, Sandy and I observed and photographed MLs at the southeast edge of Mitchell Flat in a region known as Whirlwind Mesa (Story 27). These events were also recorded using the IR camera. Figures 49-51 are three selected photographs out of 152 taken that night. I sent a larger selection of photographs to my brother, Will, for his review and comment. Even though he is half a continent away (Coos Bay, OR), he is an active contributor to my project via email. We have ongoing and frequent discussions. Below is an abbreviated version of his response to the pictures I sent him. It contains technical jargon typical of many of our exchanges, but I include this only as a sampling of our email exchanges:

Sent: Sunday, October 29, 2006 9:35 PM
To: James Bunnell

Re: October 19, 2006 ML details

Hi Jim —
This display is outlandish. Can't think of any word to describe it. No phenomena or any mix thereof that I know about could produce such effects.

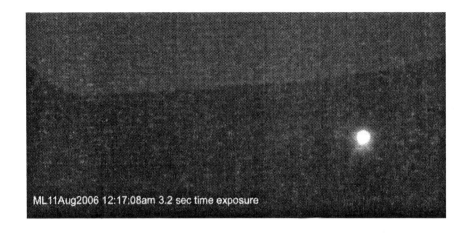

ML11Aug2006 12:17:08am 3.2 sec time exposure

Figure 47. ML of 11 August 2006 was moving left and right but stood still long enough for this 3.2 second time exposure.

ML11Aug2006 12:17:24am 5 sec time exposure

Figure 48. ML of 11 August 2006 moved right, left and right again. This five-second time exposure was taken as the ML decayed just before going out.

103

As you note, there is a strong tendency to form horizontal extensions or layers. The Earth's magnetic field dips at about 60 degrees in that area, so, other than the occasional "descending stream" your photos capture, the Earth's field doesn't seem to be playing a big role.

Horizontal extensions or layers, (I kept counting four layers, sometimes the upper layers or extensions appear to be twined or doubled) imply as a first guess something to do with atmospheric pressure, or possibly temperature gradients. Do you know what the wind was doing during this display?

It might be possible that an electric charge field is involved, as if the tuff layer had built up and sustained an electric charge — this would diminish or lapse linearly as you move away from (or upward from) the charged surface, provided that the charged surface is large in comparison to the distance you are from it. Otherwise, the inverse square law intervenes.

Since the shape and gradient of the electric charge field would depend on the shape of the charged surface, then irregularities in the tuff layer (if that is where the charge is being maintained) might account for the shapes of some of the pointy halos visible in some of the photos. Possible, but just barely.

Ordinarily, an electric field exerts its influence by the attraction of opposite charges, or repulsion of unlike charges. Were the tuff field the culprit, and held, say a negative charge, then it could in theory attract a positively charged ion cloud, and stability could be achieved by balancing that attraction against thermal lift of the ion

104

Figure 49. 19 October 2006 ML photographed with an infrared (IR) capable camera while moving right.

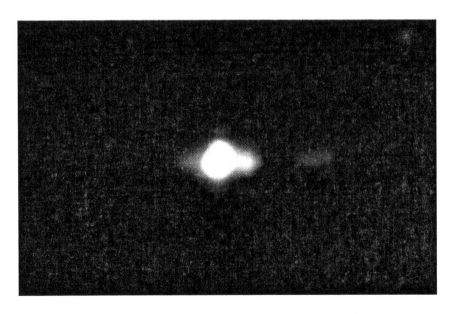

Figure 50. IR picture of ML moving left during a second appearance on the same night.

105

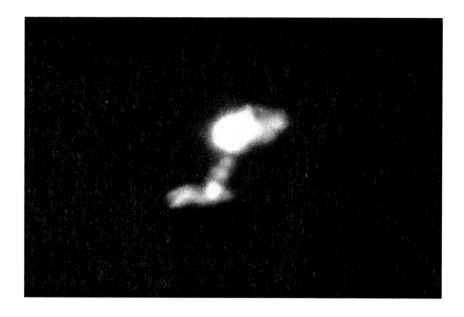

Figure 51. At 10:02 PM CST on 19 October 2006 the ML executed an odd vertical movement downward as captured by this time exposure. It then returned to a more normal circular shape and continued left until going out.

cloud. Again, possible, but just barely. Wind would probably mess this up thoroughly.

Meanwhile, some form of energy conversion is taking place, to generate and maintain the visible ML. What you are seeing is photons released when electrons change orbit or level. All visible light is that, without exception. Energy must be supplied to force the electrons to change orbits; when they fall back, they release that energy as photons, which we see.

So where is the ML energy coming from? What energy transformation is taking place? As you say, combustion, perhaps methane undergoing oxidation is the culprit. In any case, a good

106

spectrogram of the ML would tell much. I mean where you have a slit, diffraction grating or prism, and all the required optics. A spectral diffraction of the whole, un-silted image of the ML is just too broad or smeared out to tell a whole lot.

Will has a keen intellect and he always contributes an interesting perspective. As he comments at the end of his message, having a real spectrometer rather than just broad spectral displays produced by diffraction grating filters might go a long way toward supplying real clues to ML constituents. All of my earlier attempts to find a suitable spectrometer were unsuccessful. Spectrometers designed for laboratory or student experiments tended to be inadequate. Professional spectrometers and astronomy type spectrometers were all too big, too elaborate and too expensive. But, in the following year finding a spectrometer would become a lot more feasible.

Infrared ML -- July 2007

My infrared (IR) camera would find another ML target on July 23, 2007, that also revealed interesting IR features as shown in Figure 52.

Burning MLs -- July 2008

On a warm night in July 2008, Snoopy captured burning ML activity that lasted 27 minutes and 38 seconds. This ML was estimated to have been located approximately two and a quarter miles from Snoopy. The video clearly shows what appears to be expansive-dynamic flame activity for the primary ML from birth to finish along with brief appearances of a smaller short-lived companion. Shown in Figure 53 is a snapshot of the primary ML and its small companion.

107

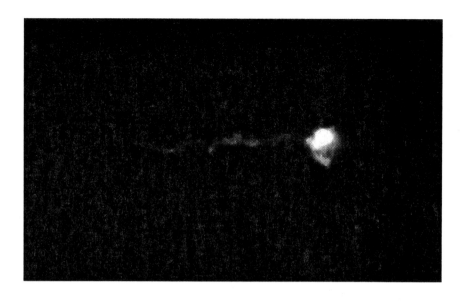

Figure 52. ML07232007 captured by infrared capable camera shows details that were not visible to the human eye.

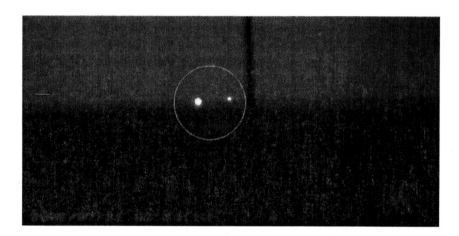

Figure 53. MLs on 30 July 2008 located approximately 2 1/4 miles from Snoopy displayed dynamic burning processes during its 27 min 38 sec lifetime.

108

"They bounced up in the sky, zoomed together, split apart and did all sorts of crazy things. At times there would be 6-7 of them bouncing around. They would start from fairly far apart, merge into one, split apart again, and then race off in any number of directions." -- from Story 20

The Search Intensifies

Measuring Magnetic Fields in Mitchell Flat

As previously mentioned, Mitchell Flat is geologically unusual in that it is completely covered with a thick layer of zeolitic-tuff that was deposited millions of years earlier. Back in 2003 I wondered if these ancient volcanic deposits might have magnetic properties because I was starting to suspect that electromagnetism might be an aspect of ML appearances. To see if the static magnetic field in Mitchell Flat was unusual, I purchased a sensitive three-dimensional magnetometer and took a total of 22 measurements along Nopal Road, an unpaved road that cuts through the middle of Mitchell Flat. The average magnetic field strength was 548.7

milligauss with a standard deviation of only 1.7 milligauss. Those results were ordinary for the Big Bend and similar to readings taken at a number of other locations within Presidio County. The conclusion of this study was clear. The magnetic field in Mitchell Flat is not unusual and therefore zeolite concentrations, to the extent that they exist in Mitchell Flat, do not appear to affect static magnetic field values.

I wondered about the possibility of temporary electromagnetic anomalies associated with ML appearances? To test that idea, I began carrying hand meters to measure changes in electromagnetic fields anytime an ML was sighted. My readings from this type of meter would be in electrical units (see Figure C11 in Appendix C). In those days I would monitor for MLs until about midnight, return to my motel in Alpine for sleep, and then rise early enough to be back at the View Park an hour or two before sunrise in hopes of catching any early morning MLs. September 13, 2003, was one such morning. As I drove across from Alpine I could see a light racing across Mitchell Flat. It was pulsing on and off as it traveled west and I felt sure this was an ML, not a ranch truck on Nopal Road. I raced on to the View Park where I hurriedly parked, jumped out, and quickly set up my camera and tripod. I had time enough to take only one clear picture before the light extinguished (see Figure 54).

Once I was convinced that the ML had gone out for good I

fished out my EMF meter and took readings:

Hrs After	Volts/Meter	Frequency	Range Measured
0.25	20	VLF	500Hz to 400 kHz
0.25	40	ELF/ULF	5Hz to 500 Hz
1	3	VLF/ELF/ULF	5Hz to 400 kHz
5.5	3	VLF/ELF/ULF	5Hz to 400 kHz
9	0.4	VLF	500Hz to 400 kHz
9	0.3	ELF/ULF	5Hz to 500 Hz

The 20 and 40 volts/meter values recorded approximately 15 minutes after the ML event ended were quite strong, especially considering that the ML had appeared to be some distance away and completely off for a quarter of an hour. Within one hour after the ML event, meter readings were clearly decaying. These recordings supported my guess that ML events would have associated electromagnetic components.

Parsing Light

The September 13, 2003, morning ML was significant in another way that was perhaps more important. I would have loved to use a spectrometer to find out more about the chemical makeup of MLs but had no such device. However there was another strategy that looked simple and relatively easy to accomplish. It is possible to purchase sheets of plastic material that are covered with thousands of vertical and horizontal grooves, creating thousands of small triangles. These materials are called diffraction (DF) gratings and they are used for the purpose of breaking down light by frequency. We have all seen how light shining on a prism emerges on the other side as a rainbow of colors. The amount that light bends

111

when striking a prism is a function of the light's frequency. Because of this fact, the prism has an ability to parse light by frequency, resulting in a rainbow of colors. That is also what causes rainbows. In the case of rainbows, rain drops act as the prism.

The practicality of DF material is that it can be easily cut to any shape. I obtained a couple of sheets of DF and experimented with placing it in front of my camera lens to discover the frequencies of photographed lights. The immediate problem was that the light and its spectra would not both fit on a single picture image. I had a choice. I could photograph the ML or its spectra but not both in one picture. I solved that problem by moving the DF filter to a location between the camera and the lens where the light cones were close to their crossover point. This scheme worked and it became possible to photograph a mystery light and at the same time capture a sampling of its spectra all in one picture.

On that fateful morning of September 13, 2003, the camera and lens I hauled out barely in time to snap one time exposure was equipped with my home cut DF filter and I did succeed in capturing the ML's spectra (Figure 54). Resultant spectra were discontinuous indicating that the ML was plasma. This result seemed to answer the question raised by the remarkable February 19, 2003 photographs (Figures 30-32). Even though the photographs taken in February had looked like chemical combustion processes, based on this spectral evidence, they must have been balls of plasma.

As previously mentioned, plasma is a fourth state of matter and what stars are made of, but we also utilize plasma for television displays and lights. Fluorescent lights, mercury vapor lights and sodium vapor lights are all examples of manmade lights that use plasma as light sources. ML displays are very energetic and that would be consistent with plasma because the plasma state

112

Figure 54. ML located due south of View Park on morning of 13 Sep. 2003. This photograph clearly shows discontinuous spectra (spectra can be found along the slanted grey line) characteristic of a plasma light source. The light was moving west so it could not have been a mercury or sodium vapor ranch light. Other MLs have shown continuous spectra consistent with non-plasma combustion. See "What Are They" chapter for discussion.

requires a high level of heat energy. The photograph taken September 13th provided unmistakable evidence of plasma because the spectra were discontinuous, an outcome possible only with plasma light sources.

The fact that this early morning ML was plasma meant that it was not vehicle lights because headlights use filaments made of solid materials that produce continuous spectra. I reasoned that if MLs are plasma, then diffraction (DF) gratings might provide an easy way to distinguish between car lights and MLs. I inserted a DF filter in the left eye piece of my main binoculars. With these binoculars I could grab a quick look at any suspect light and determine almost instantaneously if it was a plasma light or non-plasma and, therefore, not an ML (or so I thought at the time). Of course, mercury and sodium vapor ranch lights were plasma lights and do return discontinuous spectra, but they posed no concern because ranch lights are always fixed, not moving, and their locations were known. Hallelujah! My study was starting to bear fruit. Nights of wondering, "Is that an ML or only a ranch truck?" were finally at an end. By simply hoisting my modified binoculars I could determine the answer in an instant. Life was good!

For the next eight months I would use these modified binoculars in my search for MLs. During that time I did not see any unknown plasma lights, but my assumption that MLs were plasma would be shattered by MLs in May 2004, as discussed in the previous chapter. See Figures 40 to 42 for MLs displaying continuous spectra generated by a diffraction grating filter inserted between the camera and lens.

Search for Faults

How could a ground-based energy source replenish MLs traveling long distances? One possibility might be that MLs are following gaseous emissions from fault lines. That idea raised obvious questions. Where are the fault lines in Mitchell Flat and do they coincide with computed ML locations?

I reviewed all available geological books that provided coverage of West Texas regions. Some of the references did show fault lines in West Texas, but they all stopped at the edge of Mitchell Flat. I consulted with the University of Texas and obtained geological maps that reflected fault lines in Mitchell Flat and surrounding areas. Unfortunately, these references seemed to show generalized faults that were representative rather than specific, detailed locations in Mitchell Flat. These various sources were all helpful but none of them agreed and none provided the kind of specific location information I was seeking.

I next tried to find fault lines on my own using Google Earth satellite maps. This proved to be an interesting exercise but required care because cow trails, old roads and gullies could be easily mistaken for fault lines. These overhead satellite photographs did show other consistent detectable lines that generally ran in directions suggested by various geological studies. Using this source I started constructing fault line maps and then verified some of these features on the surface in Mitchell Flat.

In the middle of this exercise a new friend, James (Jimmy) Nixon III, provided me with a set of fault drawings he had made using Landsat Satellite photographs. His maps were much different from those I was constructing. The products he had used to derive fault lines were a good deal better than any at my disposal so I was inclined to believe his version was best. I transferred Jimmy's fault maps into my topography computer program and began checking triangulated ML locations and ground tracks versus Jimmy's fault maps. I was not getting any matches. Either MLs were not following fault lines or else my fault reference maps were not accurate.

It is always nice when at the end of a long exercise things do match and you can say, "Ah Ha!" This was not one of those times, but it is also true that, over time, plotted ML ground tracks have tended to fall within a fairly narrow band that runs diagonally through Mitchell Flat from the southeast corner to the northwest corner. If we take a broad perspective it must be noted that major faults, including the Walnut Creek fault, do run in the same general direction as ML ground tracks and my fault maps. There is insufficient evidence to say if MLs are tied to, or not tied to, a system of fault lines in Mitchell Flat.

Texas State University Takes an Interest

Dr. Karl Stephan, an engineering professor at Texas State University, purchased my book, *Night Orbs*, and in 2005, asked to meet with me in Austin to discuss my investigation. We soon became friends. In 2007 Karl suggested using his school's spectrometer to try and find new information regarding the makeup of Marfa Lights. Karl was already researching ball lightning and, like a number of my academic friends, wondered if there might be a relationship between ball lightning and MLs. In any case, he thought an expedition to Marfa to collect ML spectra might be a worthwhile thing to do. I was, of course, delighted with his proposal.

The scientific spectrometer Karl planned to use looked perfect for the task; I wanted one myself. Unfortunately Ocean Optics' scientific model exceeded my self-funded budget. I would have to look elsewhere. An internet search turned up an affordable spectrometer produced in Canada. The Canadian model had been designed for lab work (the usual case with this class of instrument). It had a more narrow frequency range and was less sensitive than Ocean Optics' scientific model, but it was more affordable and had adequate specifications, so I bought one. Sandy and I began working on how to best use this new device. Meanwhile, Karl made plans for a scientific expedition to Marfa in the spring of 2008.

Karl and I met in Fort Worth a couple of times to exchange information and ideas. Karl adapted the school's spectrometer to mate with a personally-owned telescope and I did the same with my new spectrometer. I could hardly wait to try out this new device in Marfa.

Sandy and I traveled to Marfa on December 17, 2007, with our new spectrometer and tried to capture spectra from an un-

116

known light on the 18th at 10:20 PM CST. There is no way to know if the light we saw actually was an ML for several reasons. Sandy was in our truck operating the computer (necessary for spectrometer capture) and for that reason was unable to see the light. I was outside managing our acquisition telescope. I could see the unknown light (UL) but all of my attention was focused on keeping the target within the spectrometer's narrow field of view as it darted across the desert (not easy to do). We did manage to collect spectral profiles, but our captures were all too weak to be accepted as real data. It was our first attempt with this spectrometer against a real unknown target and we were still at an early stage of learning what would be required to use this instrument effectively.

If I had any doubts about our collected profiles being only noise, our next trip to Marfa would dispel them. On that trip in February 2008, we saw no MLs but did learn our setup was not ready for prime time. The primary difficulty was what I called a "thread the needle" problem. The optical cable used to convey target light from the telescope to the spectrometer was only 600 micrometers (0.0024 inches) in diameter. Following that trip, I upgraded this optical conduit to a larger 1000 micrometers (0.0040 inches) optical cable, but that is still a tiny diameter. A distant UL is also an extremely small point of light. Getting two tiny diameters to merge by pushing around the tracking telescope was almost impossible for a stationary target and even more difficult if the target was moving (as most MLs do).

To deal with this alignment problem, Karl and I had both opted to use an Ocean Optics collimating lens that had capability to increase the capture basket by a small amount. The collimating lens causes the spatial cross section of incoming light to become smaller in order to focus the beam into the fiber optics cable that feeds the spectrometer. Ocean Optics' collimator increased the capture basket to approximately 5 mm. Based on analysis, Karl had concluded that the collimating lens was not large enough to be practical. To solve the problem, he developed a design that incorporated two collimating stages and thus spread the capture basket to a more reasonable size (~ 1 inch dia.). I had independently come to the same conclusion but had not taken the additional time required to redesign my setup. After our trip to Marfa where we experienced difficulty acquiring even fixed targets, it became obvious that a second collimating lens would be necessary.

I contacted Karl to learn what second collimating lens he had selected for his system. He did much better than that; he furnished drawings of his complete spectrometer adapter. His design was excellent and a credit to his engineering skills. I had a local machine shop make one for me. The second collimating lens was the most important component, but his design also used a light splitter so that he could simultaneously videotape light targets and send light to the spectrometer. It was a good scheme, so I acquired a compact digital video recorder to take advantage of that design feature.

Our telescope setups were different. Karl used an electric driven "GPS Go To" telescope mount and his colleague, William Stapleton, modified the telescopes' software to enable logging of azimuth and ascension angles along with a separate GPS time source. This gave Karl's setup complete bearing and time information for every recorded event.

For my part, I elected to go with a manual "Push To" altazimuth mount (Figure 55) because of concern that tracking moving ML targets might be difficult without freedom of movement inherent in a "Push To" telescope mount.

118

These upgrades, along with going to a larger telescope, made a difference in system performance, but lack of sensitivity is still a concern for my setup. Marfa testing showed that the improved configuration was able to capture plasma light sources (e.g., mercury vapor and sodium vapor ranch lights) at realistic distances but was severely limited for continuous light sources (e.g., burning solids, liquids and most gases). The ML photographed in 2003 showed discontinuous spectra (Figure 54), but multiple MLs photographed in May 2004 (Figures 40-42) showed continuous spectra. If MLs do have continuous spectra, then our ability to acquire them with my instrument is, unfortunately, limited to fairly close encounters.

Karl Stephan, with his graduate assistant Sagar Ghimire,

Figure 55. My spectrometer setup with a "Push-To" telescope.

began a planned twenty-night watch for MLs on May 11, 2008, using their more sensitive university spectrometer that is fully capable of acquiring line (discontinuous) or continuous spectra at realistic distances. Unfortunately, MLs are a fickle bunch and elected not to show themselves during that time frame. I was disappointed and am sure Karl and Sagar were as well. Nevertheless, the twenty-day expedition produced useful science as documented in a paper titled, **Spectroscopy Applied to Observations of Terrestrial Light Sources of Uncertain Origin,** authored by Karl D. Stephan, Sagar Ghimire, James Bunnell and William A. Stapleton. This paper was accepted by the American Journal of Physics and scheduled for publication in late 2009. The paper showed that a combination of computer azimuth and altitude logging, video recording, and continuous spectroscopy provided enough data for unequivocal identification of false positives such as automobile headlights, fires, and other explainable light sources. In addition, the paper demonstrated that spectroscopic analysis of molecular oxygen absorption can be used to determine distances of continuous-spectrum objects with an accuracy of +/- 1.4 km or better. The paper also showed that astronomical objects are useful for both directional references and light-flux calibration which allows estimation of the absolute radiant intensity of distant terrestrial objects, and thereby aids identification and location.

This was useful information that could be applied to future expeditions. As of this writing, neither Karl nor have I had an opportunity to capture an ML spectrum with either the university's spectrometer or mine. To do so will require spending significant time in the field (especially so in my case because close encounters are rare). As an alternative, Karl proposed adding diffraction grating material to one or more of my black and white monitoring cameras. Without color one might think the resultant spectra would be useless. However, it appears that even spectra from black and white images can be analyzed by computer using specialized software. To verify his idea, Karl conducted experiments using a similar black and white video camera, lens, and diffraction grating filter to extract spectra from plasma light. His results with software manipulation were consistent with the spectral profile obtained using the university's spectrometer. These results were encouraging and clearly demonstrated the concept was viable, at least for plasma lights.

Based on these preliminary results, I added diffraction grating filters to Snoopy's two cameras in December 2008. At this writing, technical details are still being sorted out, but the approach looks promising. The quest to determine the composition of MLs continues.

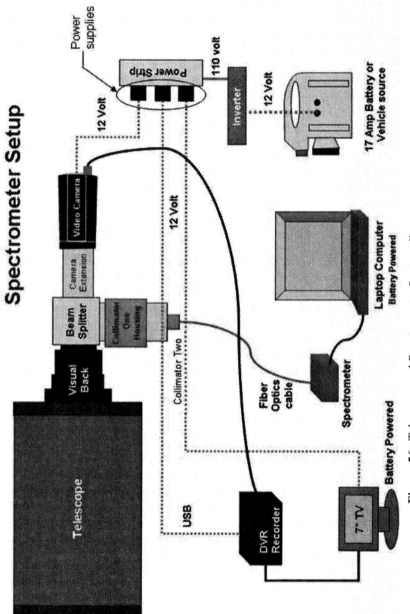

Figure 56. Telescope and Spectrometer System diagram.

122

"...the pilot pulled off his night vision goggles to get a better look but the ball of light disappeared. Able to see nothing unusual with his naked eyes, the pilot pulled his night vision goggles back into place and, presto; there was the ball of light once again." -- from Story 24

Night Discoveries

Unusual Events

My monitoring stations were photographing large chunks of Mitchell Flat every night and, as you might expect, they were finding more than just MLs. The downside: I had saddled myself with a large review task that was indeed challenging to accomplish. The upside: My monitoring stations were uncovering fascinating stuff -- I never knew what might pop into view next. Wife Sandy can attest as to the many times I have asked for her thoughts on the latest captured video oddity. Finding unusual happenings has been going on for a long time now and is a story in itself. We will dip into these strange waters briefly in this chapter to review a few of the more interesting video captures.

Weird Stuff

Roofus records strange and unknown images from time to time. I have no idea what creates these displays so I call them Weird Stuff (WS). WS events last about two minutes and, though rare, they always seem to occur in the same location. So far I have not been able to associate WS with MLs or other events although it is natural to suspect that they may be another aspect of MLs. Shown in Figures 57 & 58 are two examples from different nights in May 2007.

Another Electric Surprise

Snoopy detected an unusual electrical event involving a vehicle traveling west on US 90 at 1:50 AM CDT (0650 UT) on June 25, 2008. This was the first of three similar events, all in the same location, but on different nights. The event lasted for more than ten seconds -- too long to be a lightning bolt. Nevertheless, I suspect this event may have been related both to lightning occurring in the area at the time and to some unidentified electromagnetic anomaly (Figure 59). I would have attributed this display to optical distortion caused by water on the camera housing, but it occurred only once that evening. Other vehicles passing the same location immediately after the event were unaffected. Also, this particular vehicle showed faint dynamic residual effects extending above the vehicle as it continued west. These continuing events eliminated optical distortion as the source.

Two of these three electrical oddities involved vehicles traveling US 90; the third involved a train locomotive close to US

124

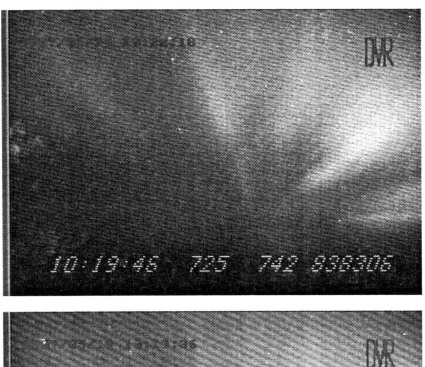

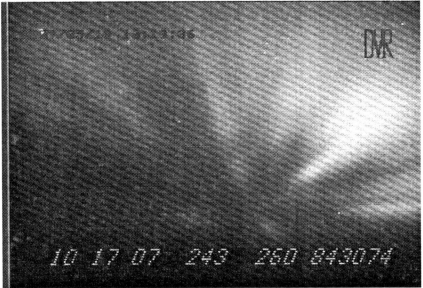

Figures 57 & 58. Weird Stuff (WS) seen by Roofus on four different nights in 2007. These events seem to fill the sky for about two minutes at a time. Are they related to MLs? We do not know.

Figure 59. This is not an ML. It is some kind of electrical event involving a motor vehicle traveling west on US 90 close to Paisano Pass. These events happened to automobiles on two different nights in the same location plus one event involving a train locomotive near the same spot.

90. These events would seem to indicate the presence of electro-magnetic anomalies near Paisano Pass. That is the same location where two remarkable car chase stories were reported to have occurred (see Stories 33 and 34).

Unusual Meteors

Meteors are commonly captured on my monitoring station videos. They vary in size and brightness. Figures 60 and 61 show the largest and brightest observed so far. This meteor appears to have exploded over Goat Mountain although meteor hunters I have consulted believe it must have been located farther away.

Figures 62 shows a meteor that appeared to be going up instead of down. It is an illusion created by a far distant meteor that

Figures 60 & 61. A large meteor descends and appears to explode over the south end of Goat Mountain. Meteor hunters say that it was probably located further away.

127

Figure 62. The video of this meteor shows it going up instead of coming down. This illusion was created by curvature of the Earth when a far distant meteor entered the atmosphere below the horizon and then rose above the horizon as it came into Snoopy's field of view.

entered the atmosphere below the horizon (because the Earth curves) and then appeared to rise above the horizon as it came into Snoopy's field of view.

Venting from Whirlwind Mesa

Figures 63 and 64 show something venting into the atmosphere at the southeast edge of Mitchell Flat (above Whirlwind Mesa) as captured by Owlbert Camera A3. Because monitoring station cameras take black and white images, it is not possible to tell if these vented substances are multicolored, but they may be curtains of light similar to those seen by Jimmy Nixon (see Story 25). This curtain-like phenomenon will be discussed in Part II.

Figure 63 & 64. Some unknown material is shown venting above Whirlwind Mesa at the southeast edge of Mitchell Flat. Similar venting has been captured by Owlbert night cameras a number of times.

Fast Movers

From time to time my cameras have photographed airborne lights moving across the sky at incredible speeds. Most of my night cameras stack images to accumulate light. The result of this image stacking is creation of time exposures equal in length to the number of stacked images. In the case of fast-moving light, these time exposures create light tracks and the length of the light track corresponds to how fast the light source is traveling. To convert the length of captured light tracks into miles per hour I have to know how far away the light is located, information not normally available, but the length of the light bar does give me a quick first estimate of speed. For example, commercial airliners have light tracks that are typically 1/8 to 1/4 inch in length. So when I see 3/4 inch light track I know the source is moving at high speed even though I may not know exactly how fast. In one case, triangulation showed an unknown, fast-moving light, to be located within 35 miles of Snoopy and traveling approximately 18,000 mph. Figures 65 and 66 show another fast- moving light that appears to have come from the direction of Mexico. The length of the light bars suggest a relative speed considerably faster than commercial jet traffic and well above the speed of sound. An astronomer friend reviewed some of these photographs and suggested that they are most likely meteors. He is probably right about that, but I am not convinced that explanation is adequate to explain all recorded high speed flyovers. I still find them interesting and puzzling.

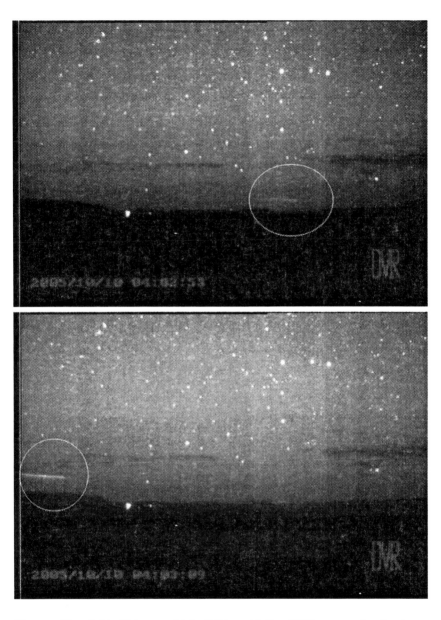

Figures 65 and 66. Unknown Light (UL) on 10 Oct. 2005 crosses the sky three to four times faster than commercial jets. Fast movers like the one in these photographs are captured by night monitoring stations every now and then. Meteors? Reentering space debris? Maybe. All we know for sure is that they are too fast for conventional aircraft. This one appears to have come out of Mexican airspace.

Unusual Lightning Events

Figure 67 was taken by Roofus on June 13, 2003. It looked like a high altitude balloon but I suspected that it was actually an unusual type of lightning phenomena called a sprite. About two decades ago, airline and military pilots were seeing sprites above electrical storms and wondered what they were. Investigation by weather people confirmed that they were unusual lightning events. At first it was believed that they were being emitted by intense electrical storms and going up into space. Satellite radar would later reveal that they actually originate at the edge of space (60 miles high) and descend down into thunderstorms. They were also found to occur in connection with positive lightning bolts. The majority of lightning bolts are negative but the most powerful ones are positively charged.

Figure 67. Unusual UL on 06/13/03 turned out to be a type of lightning known as a sprite. This sprite is located above a storm that is out of sight below the horizon. Top of the sprite is at the edge of space. This photograph and another one were made possible by illumination from a full moon.

Sprites are very short duration events and are obviously located very high in the sky (tops of sprites are always at the edge of our atmosphere). It was a little hard to believe that my Roofus Sony camera might have been able to capture such an exotic event that, calculations showed, would have been located about three hundred miles away. I located a national expert on the subject, Dr. Mark Stanley, and sent him two photographs in January 2004 to inquire if they might be sprites. He replied that they were indeed sprites. I would later realize that it was the presence of the full moon that provided enough illumination for my Sony camera to grab these images.

It was the only time that camera was able to capture sprites, but after upgrading to more light sensitive cameras, detection of sprites became common place during thunderstorm season. Figures 68 and 69 show examples of sprites from September 2006.

A Giant Jet

There are even rarer types of lightning events called "blue jets" or "giant jets," depending on size. Jets are discharges that go upward into space. They look something like a fireman's hose with branches at the top as they start to exit the Earth's atmosphere.

These powerful discharges are seen above storm tops and last about 80 milliseconds. Around 11:30 PM on May 12, 2005, both Roofus and Snoopy managed to capture clear pictures of a giant jet. The jet was followed by an extensive display of sprites in the morning hours of May 13, 2005. Roofus' picture of this giant jet is shown in Figure 70.

After posting this picture on my website, I received an E-mail response from a French scientist, Oscar A. van der Velde, who asked if I knew how rare this photograph was. It turned out that

Figures 68 & 69. Example sprites from Sept. 2005. Like snowflakes, sprites are all different. Sprites occur above distant storms that do not show in my photographs because they are so far away as to be located below the horizon.

134

photographs of any jet are very rare and photographs of giant jets over land are almost non-existent. Thanks to Oscar and an American weather scientist, Dr. Walter Lyons, I would learn that Roofus and Snoopy had managed to capture the first giant jet photographed over North America. This unique find was published in the *Journal of Geophysical Research, Vol. 112, D20104, doi:10.1029/2007JD008575, 2007* titled, **Analysis of the first gigantic jet recorded over continental North America,** by Oscar A. van der Velde, Walter A Lyons, Thomas E. Nelson, Steven A. Cummer, Jingbo Li and James Bunnell. Our study showed that the jet had been located in Mexico south of Del Rio, Texas. It had indeed been an unexpected and valuable photographic capture.

Figure 70. Giant jet photographed by Roofus on May 12, 2005.

"It was obvious that this owl was flying in tight circles so he could look through the windshield of our vehicle at us. He was curious about us."--from Owl Intelligence (this chapter)

Creatures and Critters

No story of hunting light would be complete without some mention of wildlife living in Mitchell Flat.

Bugs of Light

For a light hunter, the most significant creature of the night is an insect (Sandy's response to this statement was, "Baloney! It's rattlesnakes!"). Some of my early ML photographs contained small, short, blue light tracks of unexplained origin that could appear anywhere in the camera view, either above or below the skyline. The amazing source of these light tracks turned out to be a small bug that emits a blue light from his tail. The one I found at the View Park in June 2003 was in the larva stage and looked like a chain of small tapering beads with alternating colors. The light was located in the very tip of his tail, which he would curl down and

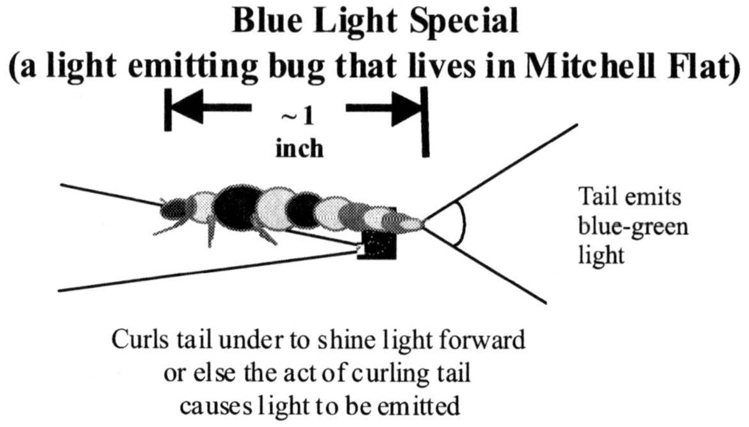

Blue Light Special
(a light emitting bug that lives in Mitchell Flat)

|←— ~1 inch —→|

Tail emits
blue-green
light

Curls tail under to shine light forward
or else the act of curling tail
causes light to be emitted

Figure 71. Some lightning bugs in Mitchell Flat emit blue light.

forward to point the light between his legs. I do not know if this curling action was to shine the light ahead or was necessary to produce the light. In any case, he was creating a blue circle of light about 6 to 8 inches in diameter. He turned out to be a nocturnal luminous insect belonging to the beetle family, Lampyridae -- a type of lightning bug. There are over 2000 known types of light-emitting bugs and this was one of them! In a later stage, this little bug is capable of flying, creating blue light tracks in the sky, and leaving them on some of my photographs.

My monitoring station cameras see these bugs as white lights and they create displays that can be mistaken for MLs. With experience, I have become more skilled at filtering out these potential sources of data contamination.

Social Coyotes

Roundly despised by ranchers, these predators have a rough

life: finding enough food and water every day to survive, as well as avoiding poison, traps, and gun-toting ranchers. Of course, coyotes are killers who would just as soon grab a small calf as a rabbit, so ranchers have reason for wanting to do them in, but I can only marvel at their amazing ability to survive in such a hostile environment. They, of course, create a hostile climate for other living creatures in ranch country.

What I learned in Mitchell Flat about coyotes is that they are social animals. Unlike wolves, coyotes are nearly always seen hunting alone, but it would be a mistake to think of them as being non-social. One night, sitting in the dark somewhere deep in Mitchell Flat, I realized that I could hear an animal in the brush. I could not see this animal but could hear his approach because each step was accompanied by a series of howls and barks clearly revealing his position. My guess is that those howls, barks and other sounds were not for my benefit. He was sending messages to his fellow coyotes located further away. In response to his communiqués, every other coyote within miles would respond in kind and then wait for the next message. I guess he was telling them how he was doing and whether I was a potential meal or a serious threat.

Listening to this extensive ongoing communication between these wild animals was fascinating; it was clear that these sounds were not mindless barks, they were language. Those coyotes were talking to each other. About that I have no doubt, and it was amazing.

At no time did I feel frightened. Coyotes do not normally attack people; they fear people and with good reason. My guess is that once this coyote determined that I was human, he retreated and the barking exchange was over.

Owl Intelligence

Rancher Kerr Mitchell genuinely appreciates animal life and takes an active interest in the many wild and domestic creatures of his domain. He called to my attention a family of burrowing owls that had taken up residence on his ranch land. These small owls don't bother to build above-ground nests, but rather inhabit ground dwellings constructed by other desert creatures. They are cute little critters and Kerr asked me to use my long range lenses to photograph them. It was a fun assignment. Sandy and I spent a the better part of a summer day trying to sneak close enough to these little owls to capture decent pictures. They were cute but shy and would flee when we approached. We took the best pictures we could.

Figure 72. Burrowing Owl

That night we elected to watch for MLs from Mitchell Flat. After setting up the cameras we returned to the truck to wait for darkness. While waiting, an owl appeared and began circling the truck. He was not a burrowing owl, but I was impressed with his curiosity. It was obvious that this owl was flying in tight circles so he could look through the windshield of our

140

vehicle at us. He was curious about us. I took this as a sign of owl intelligence and commented to Sandy about his interesting behavior.

We saw no MLs that night and gave up the hunt close to midnight. As we were driving out, a small burrowing owl, with a strong family resemblance to those we had stalked earlier in the day, flew over our truck and landed beside the road ahead of us. He stood beside the road, watching us approach. I slowed the truck and stopped so Sandy could get his picture in the darkness with a point and shoot digital camera. We did not expect the picture to come out without using a flash, but it was worth a try. The resultant picture (Figure 73) is dark but recognizable; the dark area in the foreground is the hood of my truck. The owl allowed this activity

and continued to stare in our direction as I drove closer, trying to give Sandy the best possible opportunity to take his picture. When we got too close, the owl flew and then landed on the side of the road again where he resumed watching us. He was as interested in us as we were in him! We repeated the entire process several times. When we came too close, the owl once

Figure 73. Mutual Watching

141

again moved further down the road and resumed observing us while Sandy and I were doing our best to take his picture. We were charmed and impressed by this obviously intelligent little creature. Eventually he grew tired of these strange people and their bright lights and flew on into the night leaving us with big smiles and greater appreciation of these small wild creatures.

Silcock and the Owl

In October 2003, I received a nice letter from Mr. Fred Silcock of Victoria, Australia. Mr. Silcock was writing with regard to the Min Min Lights that occur in Queensland, Australia. I had read Maureen Kozicka's book, **The Mystery of the Min Min Light** and had done additional internet research on these lights because of their reported similarity to Marfa Lights. Based on what I had read, I suspected that the Min Min and Marfa Lights might indeed be similar phenomena.

Mr. Silcock's letter brought news that he had found the answer already and had written his own book on the subject. I was more than a little interested in what he had to say, but his letter did not reveal any conclusions regarding the origin of Min Min Lights. The letter did reveal the fact that he had sent a copy of his book to the Marfa Public Library.

On my very next trip to Marfa I went to the public library and asked to see Mr. Silcock's book. I sat down with the book and did a quick scan. The gist of Mr. Silcock's theory was that luminous owls were responsible for Min Min Lights! In his letter he had asked about owls in the Marfa area and, of course, there are owls in Mitchell Flat. I was unable to buy his theory but did get a chuckle at the originality of his idea.

We do inhabit a small world indeed and sometimes coincidences can occur that seem almost uncanny. That very night would find me sitting in the middle of Mitchell Flat thinking about Silcock's theory. It was a dark night and, in typical Marfa fashion, chilly with a cold breeze blowing. I was sitting in a lawn chair with a warm hood covering my head. In the dark I must have looked like a fence post because I glanced up just in time to see an owl descending rapidly toward my head with two powerful sets of claws outstretched, to grab my head on landing. The owl must have been every bit as startled and frightened as was I, because at that very moment he began beating his wings furiously in an effort to kill his decent, while mercifully retracting his powerful claws. The latter action was all that saved me from being struck; it all happened so fast, there was no time to duck. The owl cleared my head with only an inch or two to spare and disappeared into the black night in the blink of an eye. I was left sitting there stunned, breathing heavily and thankful that he had aborted his landing.

Upon my return home I would write Silcock and tell him, "Please be advised that I have sighted your owl, but he forgot to turn on his landing lights!"

Figure 74. Three owls turning final for landing at Owlbert.

Part II
What Are Mystery Lights?

"Three balls of light seemed to rise out of the ground and shoot high into the sky with incredible speed. Near the top of this display the light balls seemed to branch out into a widening pattern until they extinguished." -- From Story 30

What Are Mystery Lights?

Light Sources in Mitchell Flat

If we define "Mystery Lights" to be lights that to a casual observer have an unknown origin, then we must start by acknowledging the importance of a number of "mysterious looking" but definitely not "unknown" light sources. All of these can be sorted out, with good observation tools, time, some reference materials, and much patience. But these lights are "MLs" to many travelers who see them from the View Park, and this account would not be complete without a discussion of them.

Vehicle Lights — The most prevalent of these "mysterious looking" MLs are created by automobile traffic seen nightly traveling Highway 67 from Presidio to Marfa. People stop at the Marfa Lights View Park expecting to see, or at least hoping to see, mysterious lights in the night. Usually there is no preparation and only a few instructive signs to aid the uninitiated. Standing there peering into the dark, people see (popping into and out of view) distant lights that seem on a dark night to be suspended in the sky. These vehicles are actually on a high, winding mountain road when they first come into view. These lights do indeed look mysterious because they seem to turn on and off (the terrain masks lights as the road winds downward), vary in intensity (caused by changes in better or worse alignment with the View Park), and even do splits and merges (caused by multiple cars, trucks, or SUVs driving a curving roadway). If you too have looked southwest from the View Park in the direction of those automobile lights and believed them to be mysterious, don't feel foolish; you have lots of company. It is entirely understandable that people are going to be drawn to those vehicle lights and assume they are the fabled mystery lights.

During the month of May, 2004, members of the University of Texas Dallas Chapter of the Society of Physics Students spent four nights in Marfa conducting a grant-funded study to determine if vehicle traffic on Highway 67 might be the explanation for Marfa Lights. Their experiment included collecting video of lights located southwest of the View Park and measurements of traffic volume on Highway 67 along with use of lasers and chase cars to determine if the video recorded lights were automobiles on Highway 67. Study results confirmed that the lights observed southwest of the View Park during those four nights were automobiles. Some people believe these results constitute scientific proof that Marfa Lights are nothing more than vehicle lights on Highway 67, but that is not the case. What the student study actually confirmed was that vehicle lights on Highway 67 can be clearly seen from the View Park and that they do look mysterious.

Ranch Lights — While most View Park visitors lock onto car lights, many find other interesting, but explainable, light sources. Even fixed ranch lights look mysterious to many people. On your next visit to the View Park try this: Pick what you assume is a ranch light and stare at it. You might be amazed to find that it will soon start to move, especially on a dark night. On many occasions I have had people screaming at me, "No! It is not a ranch light! It is moving! Look, it just jumped straight up! That is not a ranch light!"

This perceived dance of stationary lights is known as the "autokinetic" effect. It is an illusion of motion that has intrigued scientists for centuries. Airplane pilots are familiar with this phenomenon since it presents, especially when flying formation, a hazard in night flights (i.e., incorrectly perceiving a change in another plane's relative position). Explanations of this include the nature of the automatic horizontal scanning mechanism of our eyes, over-compensation by the brain for eye cell fatigue and various characteristics of the light itself. You can check for reality versus illusion by lining up the light with a fixed piece of structure. As soon as you do, all sense of motion stops with a truly stationary light, and you can once again see that it is fixed and shining in a dark background.

This bit of mental gymnastics is not limited to fixed lights. Many times people assert that the automobile lights they are so intently observing cannot possibly be attached to cars because they just saw those lights fly in a circle or perform some other distinctly non-car maneuver. "Didn't you see it do that?" they ask. No, I did not, but then, neither did they.

Stars – If you have not spent time outdoors looking at the night sky you might be surprised by how interesting some stars can be when they are close to the horizon. A few of the brighter stars will seem to flash in alternating colors, an effect caused by dust and moisture plus temperature variations in the air. I am talking about much more than "star twinkling." The atmospheric effect for some stars looks more like bright alternating displays of red, blue, and white lights. These displays can be very compelling, but the effect usually diminishes and goes away completely as stars rise higher into the celestial sky. Gradually it will become apparent that you are tracking a star because the light's movement across the sky will keep pace with other stars as the Earth rotates further into the night.

Meteors – Most nights in Mitchell Flat, one or more meteors can be observed burning up in our atmosphere. During meteor showers, you may see one every few minutes and some may look quite spectacular. People unaccustomed to seeing meteors may find the West Texas meteor show (and on some nights it really is a show) unusual and wonder if these meteors are what people are calling Marfa Lights. They are not.

Iridium Satellites – This is a good one. Families of relatively small communication satellites (called Iridium), have highly reflective aluminum flat plates treated with silver-coated Teflon for thermal control. There are hundreds of these communication satellites orbiting at 780 km (~ 488 miles) altitude. They normally have a brightness of +6 magnitude requiring binoculars to even see them. But when panel angles are just right, sunlight reflecting off of those silver plates can cause them to become 30 times brighter than the planet Venus. That eye-catching brightness can last from 5 to 20 seconds before the satellite once again becomes invisible to the naked eye. It might be pitch dark at the View Park, but the sun has not yet set at Iridium altitudes and sun light glinting off of Iridium satellite solar panels may be mistaken for mystery lights. My nightly monitoring stations routinely capture these satellite passes, and those saved images certainly do look mysterious. They look like a light that approaches and then reverses direction. That reversal of direction is actually an illusion caused by variable intensity of the reflection as the satellite passes somewhere over-head. They do look mysterious but they are, in fact, another explainable light source, not Marfa Mystery Lights.

Low Flying Aircraft – The military have low-level training routes that are clearly visible from the View Park. Typical flight profiles involve aircraft flying south on the west side of Chinati Mountain, then turning east (to fly south of Mitchell Flat) and finally turning north or northeast where they can be seen skimming along, barely above and sometimes below, distant mesas at the east edge of Mitchell Flat. These very low-level flights frequently go unnoticed, until they are seen skimming along about even with mesas southeast of the View Park. Are these a source of some mystery light reports? Most likely they are.

Lightning – In summer months Mitchell Flat is home to many thunderstorms and associated electrical discharges can be spectacular. As previously discussed, storms located so far away that they are below the horizon can sometimes produce enough lightning to illuminate the night sky. When storms are that distant and the thunder goes unheard, it can be a little unnerving to see the sky light up from an unknown "mysterious" source.

Sprites – A special form of lightning called sprites (see **Night Discoveries** chapter and Figures 67-69) can be seen above distant thunderstorms if you are looking in the right direction. They appear as eye-catching columns of red filament high in the sky. They are so brief as to leave the viewer unsure of what he or she has seen, yet aware that it was something unusual. Most people have never heard of or seen this phenomenon, so such observations are likely to generate mystery light reports.

Fireworks – Who would expect to see fireworks shooting into the sky from somewhere out there in Mitchell Flat? My night cameras have seen such human-created displays three or four times, especially on or close to July 4th and January 1st.

Train Lights – Trains have lights brighter than automobiles and from a long distance, they can look unusual. Active train traffic is parallel to US 90 and normally easy to recognize if you are at the View Park. But looking due west, you might spot lights from trains on their way into Marfa from the west. In that direction the lights can be far enough away to look mysterious. If you are using binoculars, you may see evidence of red, yellow or green lights that change colors from time to time. These are railroad section lights that can also be seen at closer range east and west of the View Park.

There are also train tracks that pass through the middle of Mitchell Flat and go south to the Mexican Border. When running, occasional night trains and associated section lights become another source of mystery light reports. (The Rio Grande railway bridge between Presidio and Mexico burned in 2008, stopping train traffic in and out of Mexico pending repairs)

USAF Radar Blimp – An aerial surveillance radar blimp is located to the west, about halfway between Marfa and Valentine, near US 90 (see Figures 75 and 76). The purpose of this aerial radar is to monitor air traffic entering the United States from Mexico. The blimp can be easily seen during daylight hours and is

Figure 75. The USAF flies a surveillance radar blimp from a station located between Marfa and Valentine to detect illegal air traffic entry from Mexico. The blimp is tethered with a steel cable and is able to fly as high as 10,000 feet, making it clearly visible west of Marfa during the day and as a flashing light at night.

Figure 76. What the blimp looks like up close.

153

sometimes the source of mistaken UFO reports. It is equipped with aircraft clearance lights and, at night, those flashing lights hanging high in the sky are sometimes reported as mystery lights.

Lightning Bugs – Those cute little critters are more than happy to put on impromptu light displays in white or beautiful blue color. How many blue lightning bugs have most people seen? There are some in Mitchell Flat and it is easy to understand how these light sources are sometimes reported as mystery lights flying in the night. They are also captured from time to time by my night surveillance cameras and these light captures have potential for misidentification.

Brush Fires and Trash Fires – Both types of fires happen in Mitchell Flat with regularity and many result in mystery light reports. After all, in good weather there are many people at the View Park looking for unusual lights and distant fires do look unusual.

Oil Fires – Oil and gas exploration continues in Mitchell Flat and on surrounding mesas. Those activities, especially gas fires, often look mysterious to View Park visitors because they bob and move in the wind.

Hoax Lights – Naturally, some people cannot resist the temptation to try and spook or fool night watchers by putting on artificial light displays. The surprising thing to me has been how rarely this occurs, but it does happen. I have seen four hoaxes and they were easy to spot. Oddly enough, I have always appreciated such efforts. Sitting out there in the dark gets boring, and foolish antics with powerful lights are worth a giggle or two. One night an oil well crew spotted me sitting in the dark, not too far from their location. They either made inquiries or else figured out why I was

154

there because the next night, after their work shift ended, they drove some distance away and then tried their best to create a little excitement by waving around what must have been a million candle power battery-powered light. It really was fun to watch and I appreciated their efforts at entertainment, even if they were pretty obvious.

Some of the people who send me their stories start by wondering if they were being hoaxed. The story they have to tell may suggest otherwise, but many folks who witness unusual ML displays would rather believe someone is hoaxing them than the alternative – that what they saw actually was mysterious, unexplained, and out of the ordinary. Many people have great difficulty accepting the idea that there might still be unknown happenings in our modern "know it all" world. Believe me folks, we do not know it all, and no one should be surprised at that revelation.

Me and We -- Believing in full disclosure, I should include myself as yet another source of mystery lights even though that is never my intention. We make eight to twelve trips a year to Marfa to collect data and to maintain monitoring station equipment. These trips provide opportunities for personal nighttime surveillance that often requires use of flashlights. Observation nights typically last from before sunset to around midnight. In order to preserve night vision, Sandy and I use red lights to see what we are doing. I turn cameras on and off, remove or insert lens covers, track targets of interest, change batteries, and so forth. At night's end, we disassemble camera/telescope combinations, take down tripods, put away tables and other equipment and check to see if anything has fallen to the ground or been left behind. The latter activity is often done with white lights as night vision no longer needs to be preserved.* Use of our spectrometer complicates our setup even more. All of this creates a number of additional light sources in the field and makes for a longer take-down phase.

Seen from the View Park, these activities may look mysterious to many observers. One night, after putting away equipment, I drove down a ranch road sometime after midnight. As I approached US 90 I could see the lights of a vehicle parked on US 90 in the direction I was going. I guessed that it was the Border Patrol, as they keep a pretty close watch on late night activities, but it was not. When I reached the highway intersection, this vehicle started up and slowly approached, forcing me to wait while he ever so slowly drove by, temporarily blocking my entry onto the highway. He was driving an old car and he leaned well out of the vehicle as he went by, straining to get a better look at me and my vehicle.

I could well imagine this man as he might have been earlier at the View Park, standing back from the crowd, listening as others were getting excited by the mysterious lights I had unavoidably made as I packed up my equipment at the conclusion of my watch. He might have suspected a deliberate hoax and driven to the intersection of the ranch road to confirm that I was nothing more than some jerk in the night, waving a flashlight around in the dark to impress tourists. I drove on to my motel and he went his way, probably to tell others how he had finally gotten to the bottom of this Marfa Lights nonsense. "It is some guy they pay to go out there and wave around a flashlight. I know because I saw him!"

*Hearing or seeing a rattlesnake also gets white lights on and probing the darkness. In warm weather, rattlesnakes do their hunting at night and to be bitten can be not only life-threatening but life-changing if you survive. The snake population in Mitchell Flat now includes Mojave Rattlesnakes -- vipers with venom powerful enough to kill cows and horses. Snakes are serious threats that must be taken into account anytime you are on the ground in Mitchell Flat.

156

And he was not the only one I have encountered late at night waiting at that intersection to get a good look at me and my vehicle. To paraphrase a popular saying, "We have spotted Marfa Lights and they are us!" -- unfortunately.

Mysterious Lights in Mitchell Flat

All of the above light sources can be, and sometimes are, reported as Marfa Lights, with automobile lights on Highway 67 and ranch trucks in Mitchell Flat being the most common. When skeptics charge that Marfa Lights are only car lights, we should admit that they are at least partly correct. The truth is that there are many sources of light visible from the View Park and most visitors manage to find something that looks mysterious to them even though additional investigation, if they had the time to do it, might reveal the source (of those unknown lights) to be mundane in most cases. Nevertheless, some lights seen in Mitchell Flat -- the truly interesting ones -- are not mundane. Based on my investigation, I conclude that mysterious lights also have multiple sources. To help keep these different light sources separate, I will refer to them as ML "types." I will list them here and then discuss each type to the extent of my knowledge about them:

1) Type M = Night Mirages
2) Type LC = Light Curtains
3) Type CE = Chemical-Electromagnetic (divided into four subtypes)

Type M MLs: Night Mirages.

Anyone who has ever driven in West Texas is familiar with water mirages on the road. Ordinary paved roads in the warm weather will sometimes look as if they are covered by a layer of water somewhere ahead of you. These "water" mirages look very real, but they disappear when your viewing angle (as you approach) becomes too steep to maintain the illusion.

Mirages are recognized perceptual phenomena and they are not limited to illusionary "water" on the road. Sometimes fence lines, telephone poles, and even city buildings will appear to be located where they are not. The more complex mirages are called Fata Morgana,[5] a name that refers to mirages characterized by multiple distortions, generally in the vertical, such that features of actual physical entities are displaced, and may be elongated or upside down.

Mirages are well-known visual phenomena, but do they occur at night and could they be a source of ML reports? The answer is yes! Mirages do occur in the Marfa Basin during the day but are more common at night for explainable reasons. The Marfa Basin is almost a mile high. Temperatures can be hot during the day but plunge quickly after sunset, resulting in cool air near ground level. The wind blows a lot in West Texas and these winds sometimes bring in, from lower level deserts, warmer air that is forced up over mountains that rim the basin. This process causes the warmer air to locate above cooler air that is trapped at a lower level within the basin, a circumstance referred to as a "temperature inversion." As a result, we have atmospheric temperature increasing with altitude instead of decreasing, as would be the normal case. Temperature inversions create atmospheric "lenses" that can bend light rays and create mirage lights that seem to be located where they are not.

I love nights in Marfa when mirage conditions prevail because they create an air of mystery. On those fairly rare nights, it seems like anything is possible and the mirages themselves are fun to watch -- sometimes even breathtaking. The first clue that mi-

158

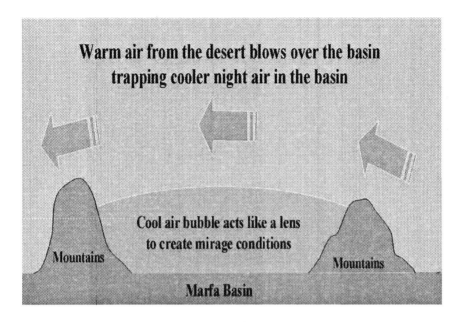

Figure 77. Temperature inversions (warmer air on top of cooler air) are somewhat common in the Marfa Basin and can result in night mirages.

rages are forming usually comes in the form of stacking and/or elongation of lights (see Figures 78-82 for examples). The last time Sandy and I witnessed mirage conditions, we were treated to seeing the lights of vehicles on US 90 "driving" through the mountains at the northeast end of Mitchell Flat, in a place where US 90 has never been. Later we would see traffic on US 90 stacked two and three deep (i.e., each car had one or two duplicates above it). The most outstanding mirage I have witnessed involved the Marfa Municipal Airport beacon. I was amazed to see it spiraling up and back down again in a repeating pattern.

When fixed lights stack, there will be one correctly placed light (usually either at top or at the bottom of the stack). All of the other lights in the stack are actually the same light seen via a curved light path caused by the temperature inversion. Moving toward or away from the actual light source can cause its mirage

159

Figure 78. These lights located west of the View Park are being "pulled-up" by mirage conditions, Nov. 2002.

Figure 79. Marfa city lights elongated by mirage conditions on 20 Jan. 2003. Night mirages do occur from time to time in Mitchell Flat and are no doubt the source of some ML reports.

160

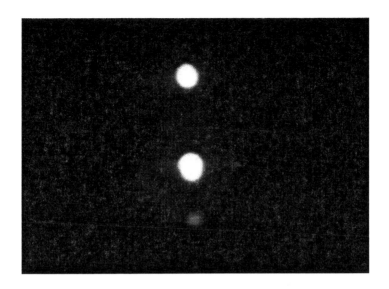

Figure 80. Mercury vapor (MV) ranch light in mirage conditions, 21 Dec. 2007. Only the top light is correctly placed. The other two are mirages.

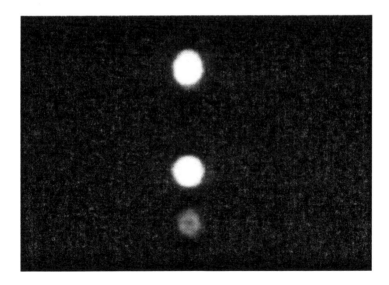

Figure 81. Later the bottom mirage light changed from blue to green. Following that, a fourth light appeared where there should have been only one.

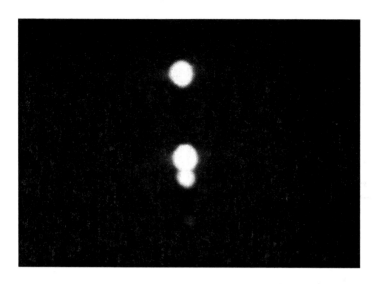

Figure 82. The fourth light seems to emerge from the second mirage light. That same night, we observed car lights on US 90 stacked three deep vertically as well as car lights moving through the mountains where no road exists.

duplicates to appear to move because the observer's movement changes his or her position within the parallel light tunnel(s).

Four ML Mirage Stories?--Night mirages might be the source of four of the mystery light stories presented in Appendix A. (You may want to skip ahead and read Stories 32, 33, 34, and 21 in Appendix A at this point, or "hang loose" for this discussion of them as potential mirage events.)

In Story 32, I suspect it was a Fata Morgana light mirage that Rob Grotty saw. He commented that the light was similar in size and the same color as the railroad section light he could see behind him and further down the track. My best guess would be that the green light he saw following him above the railroad tracks was a mirage of the very same light. As he drove at 65 mph, it seemed to be gaining on him because his position in the mirage "light tunnel" was changing. When he slowed, it slowed, and when he stopped, it

162

stopped. When he turned his truck around, he moved out of the light tunnel and the mirage disappeared.

The same explanation may also apply to Story 33, "Chased by a White Light." What Lydia Quiroz and others in the car with her saw that night might have been a mirage projection of car lights or of a fixed light located physically far from them. The mirage effect could have made the light appear close to them and chasing their car. The car that passed them was also within the mirage light tunnel and they may have perceived a stunning light source momentarily. Fata Morgana mirages are known to elongate in the vertical and Lydia described a light that was an ellipsoid with the long axis being vertical. The light disappeared from view at a point where the road bends to the right when eastbound. They may have reached the end of the mirage tunnel, or they might have lost the mirage as their vehicle changed direction with the bend in the road.

Another account of a car chase on US 90 (Story 34) involved Ms. Linda Armstrong. Could that event have also been a mirage? It is possible that it was. As luck would have it, I had one camera looking in the direction of US 90 during the time of the event. Linda described how an unknown light rushed at her car from behind, flew over or through it, and continued on westward, while climbing to a higher altitude. I reviewed the camera coverage and believe I was able to locate her car (it was one of two westbound cars that stopped at the View Park between 9:00 and 9:30 PM) but I could not find the light that flew over her car and on westward into the night. Perhaps that event happened before her car came within Snoopy's field of view. On the other hand, it may well have been a mirage that could not be seen from the vantage point of my monitoring station because the angle of view would have made that impossible.

163

We should also keep in mind that Snoopy has recorded unusual electrical events that happened near Paisano Pass (see Figure 59) where these two car chases (Stories 33 and 34) are reported to have occurred.

Mirages certainly can look real to those who experience them, and discovering an unknown bright light rushing at your car from behind, late at night, is a frightening experience. I was not there for any of those car chases but may have had a similar experience when I saw unexplained lights in April 2004 on Highway 67 (Story 21). Might the lights I saw that night also have been caused by night mirages? That was my first thought, but the idea was hard to accommodate because of the angular change from where the lights would have first come into view, and my position on Highway 67 (40 degrees, as shown in the illustration that accompanies the story). The fact that I saw a significant number of lights come over the hill, recede back over the hill, and stay in the northbound lane throughout the event, would seem to rule out a mirage explanation because mirage lights are sensitive to changes in viewing angle, especially angular changes as great as 40 degrees.

However, that 40-degree change in viewing angle was based on where the lights 'should' have first come into view. Thinking back on the event, I cannot remember exactly how close to Highway 67 I was when I first noticed the lights coming over the hill. Chances are it was less than 40 degrees, but still too great an angle for the mirage to hold its position in the northbound lane of Highway 67. But we must also consider the influence of expectations. When I saw those lights I did not question what they were, nor pause to analyze where they were. I immediately believed them to be motor vehicles northbound on Highway 67, so that is exactly where my mind would have placed them. My reaction was to speed forward and enter the highway ahead of them. After I entered the highway, I could see that they were clearly in my lane, directly behind me and, by that time, moving away. I do remember watching them go up the hill, over the top and out of sight.

Frankly, making the mirage concept work in this case is a stretch, but a logical one to make. Unless MLs are far stranger than even I can imagine, this event may also have been a Fata Morgana episode.

Owlbert Mirages -- In the last half of 2008, I pointed some of my monitoring station cameras in the direction of US 90 in an effort to cover previously unmonitored sections of Mitchell Flat. I soon started finding unexplained stationary lights that displayed variable intensity and on/off characteristics typical of MLs. These events were uncommon but I became suspicious when one location repeated on multiple nights. On rare occasions I have observed MLs repeat their location but only on consecutive nights (e.g., Appendix A stories 9 and 10; 17 and 18). But in late 2008, Owlbert Camera B2 recorded ML lights in the same identical location on three different nights. Analysis revealed the source to be a ranch light that was normally hidden by terrain. On nights when mirage conditions prevailed, because of temperature inversion, the light would stack vertically and mirage versions of the light were high enough to be captured by my B2 camera. Prevailing mirage conditions at the time also caused vertical elongation and duplication of vehicle lights on US 90.

Do Mirages Explain All MLs? It is fair to conclude that night mirages are a likely source of many ML reports, given the frequency of nighttime temperature inversions in Mitchell Flat. We

165

must consider the question of whether or not night mirages are a possible explanation for all mysterious and otherwise unexplained light encounters. I submit that the correct answer to that question is no, for the following reasons:

1) The November 2000 MLs (Stories 9 & 10) were seen by both Sharon Eby and myself from locations five miles apart with a wide angle of separation (on the order of 80 degrees). The same circumstance is true of the May 8, 2003 ML sighting that was witnessed by me from the View Park, as well as by Kerr Mitchell from his ranch (located seven miles from the View Park), and was also photographed by one of my monitoring cameras (see "ML Behaviors" Chapter). The viewing angles here were widely separated (in one case, on the order of 80 degrees). Atmospheric lenses cause mirages to have very restricted angles of view and, in both of these cases, it seems unlikely that we could have been viewing mirages from such widely separated locations and directions.

2) Mirage MLs are existing lights that have been transported via atmospheric lenses to appear in locations where they are not. The resultant mirage may appear upside down and distorted but it is only an optical version of an existing light. It is difficult to imagine any light source that would account for the combustion-like images captured in Figures 30 through 32. Not even a grass fire is likely to have been the source of those photographs because the recorded light tracks look more like strings of small explosions or eruptions rather than strings of flames. Also, the February 19, 2003 ML was moving at a pretty good clip and traveled for many miles while I was photographing it. If the light source I was photographing was a mirage it would have to have been a mirage of a moving light such as a vehicle. However, time exposures of vehicle lights do not look like a string of small explosions.

166

3) Snoopy video of an ML in July 2008 also showed dynamic flame-like activity suggesting a combustion type process that lasted for many minutes and ML witnesses have reported seeing "flame-like" ML interior processes (e.g., Story 28).

4) Night mirages caused by temperature inversions are characterized by elongation or repetition in the vertical plane (see Figures 78-82) but many ML episodes are characterized by symmetrical horizontal motion (see Figures 34 & 35).

5) A powerful argument for non-mirage MLs can also be found in events that occurred June 2 and 3, 2005. It would be difficult to account for the ML recorded on June 2 (see Figures 43 & 44) based on a temperature inversion and given its large size and unique halo pattern. Later that same night in the early hours of June 3rd, an even larger ML emitted enough radiant energy to illuminate clouds high overhead. This striking ML was apparently seen by military pilots flying a thousand feet or more overhead because they later returned and flew down over the ML's location (see Figures 45-46 and related story). Mirages do not produce radiant energy and could not have illuminated overhead clouds.

6) And then we have the remarkable and well-supported with multiple witnesses' story of Alton Sutter (Story 6) who actually held an ML fragment on the tip of his finger. What Sutter held was only a tiny remnant, but definitely physical and, therefore, not a mirage. Whatever he was holding dissipated completely leaving no residue.

Night mirages do exist and are, without a doubt, the source of many ML reports. But they are not the only source. There are other light sources even more fundamental and mysterious in Mitchell Flat.

Type LC MLs: Light Curtains.

In Appendix A, Story 25, Jimmy Nixon reported seeing a multicolored curtain of light rising high into the night sky. There have been similar reports from others. What Jimmy and others have described is not a sphere of light but something resembling a wall or sheet of rainbow-colored light. The location of these "light curtains" was in the southeast corner of Mitchell Flat, probably above Whirlwind or Mitchell Mesas. My night cameras have, on two or three occasions, photographed what appeared to be vertical discharges above those mesas (Figures 63 and 64) and, on at least one occasion, a trapped bank of clouds was seen rotating directly over Whirlwind Mesa for an extended period of time. They don't call it Whirlwind Mesa for nothing! Circulating air currents may act as suction to pull moisture (or other vapors) from the ground. These rising vapors may create a chemical stream that refracts moonlight into constituent frequencies, in the same basic process that produces rainbows.

This kind of sighting does truly cross the border into unknown territory and my explanation is only conjecture. Nevertheless something appears to be venting periodically from Whirlwind and Mitchell Mesas, and rainbow-colored curtains of light have been observed over the mesas. We can only speculate that these mysterious light curtains are related to more frequently reported ML phenomena. They differ from traditional MLs in that they are not spheres of light.

Type CE MLs: Chemical-Electromagnetic

Letters "CE" refer to MLs that exhibit "chemical" or "combustion-like" properties with electromagnetic attributes. These MLs are truly mysterious and do not fit into Mirage or Light

168

Curtain categories. For convenience of discussion, I divide these chemical-electromagnetic phenomena (i.e., Type CE MLs) into four sub-categories:

Subtype I: Specks –Tiny specks of light seen near the View Park in twilight.

Subtype II: Stationary — Stationary balls of light that turn on and off and sometimes multiply.

Subtype III: Traveling — Same as Type II but they travel cross-country above desert foliage and below background mesas.

Subtype IV: Sky — A higher-flying form that flies above the local skyline.

Laid out on following pages are what I have learned about the ML CE subtypes in three tables: Table 1 (Subtype I), Table 2 (Subtypes II and III), and Table 3 (Subtype IV).

Table 1. Characteristics of Type CE, Subtype I MLs (Specks).

#	Characteristics
1	Seen at View Park during twilight by multiple witnesses.
2	So small and short lived that you will think it was your imagination.
3	Appear solo, one at a time and last less than two seconds.
4	They fly into the wind instead of with it.
5	They shine only once.
6	White in color.
7	Flight path is straight and horizontal to the ground.

Table 2. Characteristics of Type CE, Subtype II (Stationary) & Subtype III (Traveling).

#	Characteristics	Comment	Reference: S=Story F=Figure A=Appendix Figure
1	Nighttime Only.	Daylight reports are rare and unconfirmed.	See text discussion.
2	Colors: Yellow, orange, red, reddish, green, blue, white or sequence of these colors.	Yellow-orange or reddish are most common colors followed by bright red.	S1-S4, S9-S12, S14, S15, S20-S21, S23, S25, S27-S34, A4
3	Turn on and off multiple times, sometimes described as "pulsing."	A consistent characteristic that causes gaps to occur in light tracks of moving MLs.	S1-S2, S4, S7, S9-S11, S15, S17-S19, S22-S23, S25, S27-S29, S31. F30-F33, F35-F39, A6
4	Splitting and merging.	Splitting is preceded by a bright flash but not all bright flashes result in splits.	S2-S5, S7, S9-S10, S13, S17, S20, S22, S30. F33-F35, F42, F51
5	Dancing, swinging, bobbing, oscillating, and other dynamics of motion.	Frequently the original brighter light will remain in place while off-spring orbit or dance around the primary.	S2-S5, S7, S12, S17, S20, S30. F34-F35, F37, F40-F42, F51, F52

170

Table 2 (continued). Characteristics of Type CE, Subtype II (Stationary) and Subtype III (Traveling).

	Characteristics	Comment	Reference: S=Story F=Figure A=Appendix Figure
6	Variable light intensity.	MLs seem to make step changes in brightness that include off, very dim, dim, bright, very bright, and extremely bright states.	S2, S4, S7, S9-S11, S13, S15, S17-S18, S22, S26-S28, S31. F30-F35, F37, F39-F42, F47-F48, F49-F53, A4, A6-A7
7	MLs are silent.	In my experience and no one has reported sound.	Personal experience and ML reports.
8	Size varies from pea size (S6) to huge (S26).	Most size references are to baseballs, oranges, softballs, and basketballs but the full range goes from pea to gigantic.	S2, S6, S26. F30-F35, F37, F40-F42, F43-F45, F47-F48, F49-F51, F53, A4, A6
9	Durations vary from seconds to as much as seven hours.	Median time is 4 minutes.	S2, S4, S9-S10, S12-S13, S17-S18, S20, S25, S26, S29, S31, B4
10	Travel speed varies from slow to an estimated 200 mph.	Most travel speeds for Subtype III MLs are in the 10-40 mph range. S18 calculated 37 mph.	S2, S4, S14, S18, S32, S34
11	Level flight or terrain following?	In most cases, flights are fairly level with some evidence of inclines, but MLs do lose altitude when moving from high terrain to lower terrain.	S2, S4, S14, S15, S18, S20-S22, S32, S33. F30-F36, F37-F38, F39-F42, F47-F48, F52. A4, A6, A7
12	Altitudes vary from about 3 ft to as much as 400 ft based on comparison to background mesas.	Background mesas are ~500 ft. high in most cases and can be used to estimate Type III ML altitudes.	S2, S12, S14, S15, S18, S20, S21, S25, S29, S32. F33-F35, F37-F38, F40-F42, F47-F48, F53, A4
13	What time of night?	MLs may appear anytime of night but most occur within the first 4 hours after sunset.	See Figure B2.
14	What time of the year is best?	Best months so far have been Jul., Mar., May, and Nov. Current database has no MLs in Apr. and only 1 ML each in Feb. and Sep. but these low numbers probably result from small sample size.	See Figure B3.

171

Table 3. Characteristics of Type CE, Subtype IV MLs (Sky).

	Characteristics	Comment	Reference: S=Story F=Figure A=Appendix Figure
1	Nighttime only.	Same as for Subtypes II and III.	S12, S14, S20, S30
2	Color: Yellow-white-orange, reddish, red, fiery red.	Subtype IV MLs may be redder than Subtypes II and III based on observational reports.	S12, S14, S20, S30
3	High speed of motion and very dynamic movements including vertical and horizontal extremes.	Climb in swooping motions, descend as if rolling down stairs, sudden right/left movements and up/down movements, bouncing motions.	S12, S14, S20, S30
4	Durations from seconds to as long as 30+ minutes.	My sighting (S14) was short but others have reported events lasting many minutes.	S12, S14, S20, S30
5	Splitting, merging, variable intensity, silent displays.	These characteristics seem to be the same as for Subtypes II and III.	S12, S14, S20, S30

Two characteristics common to Type CE MLs

Nighttime only – To the best of my knowledge, ranchers who have lived and worked in Mitchell Flat all of their lives have not reported seeing MLs during daylight hours. That fact is important because we are talking about people who generally work outdoors. If they are not seeing MLs during the day, then it is a good bet that MLs do not appear during daylight, or else rarely do. This constraint eliminates any suggested source that is also likely to occur during daylight. For example, it has been suggested that MLs may result from seismic forces crushing quartz rocks in small, unnoticed earthquakes, or perhaps earthquake precursors, resulting in a release of electrical sparks that create balls of light. That concept is

172

not viable because quartz crushing is as likely to occur during daylight hours as it is at night. Failure to see MLs during the day is not a visibility issue. We know that MLs are bright enough to be seen in daylight because they are often brighter than car lights. Armed with this nighttime-only constraint, we are permitted to dismiss many theories quickly and concentrate instead on a few of the more promising ideas.

Multiple restarts – One of the dominant Type CE characteristics is the fact that MLs turn on and off repeatedly (except subtype I), creating pulsing displays when they are standing still, and gaps in their light tracks when they are on the move. To be viable, our theory must account for the apparent ease with which MLs repeatedly ignite.

What Are Chemical Electromagnetic MLs?

Hypothesis 1: Type CE MLs are byproducts of solar storms.

To explain the nature of this hypothesis I must first say something about solar weather. The sun is a nuclear power plant that is continuously emitting visible light along with other forms of electromagnetic radiation, but it also emits streams of particles (plasma) known as solar wind. The solar wind is composed of ionized hydrogen and helium gas traveling at approximately 1-2 million mph. Earth's magnetic field fortunately deflects this powerful wind around our planet; without this, our atmosphere might well be blown off into space. Pressure of the solar wind reshapes the Earth's magnetic field into an elongated tear drop, with the sunward side pressed close to the planet and the night side extending far into space. This Earth-protective magnetic volume is known as the magnetosphere.

Solar weather experts have found that the sun goes through an eleven-year cycle of sunspot activity (i.e., solar storms) that generates variability in the solar wind. These wind shock waves impinge on the magnetosphere, which absorbs solar plasma that then streams down magnetic field lines to create the colorful displays we call Northern Lights and Southern Lights.

Do these solar storms also cause Type CE MLs?

Pro --

1) We know that solar storms are able to produce lights (i.e., Northern and Southern Lights).

2) The fact that the magnetosphere is compressed on the daylight side and very elongated on the night side may have something to do with MLs being nocturnal.

3) Solar storms cause electrical and magnetic disturbances that may play a role in generation of MLs.

Con --

The most important factors driving solar weather are known as "coronal mass ejections" (CMEs). These are solar eruptions that fling vast quantities of plasma far into space. CMEs that are Earth directed, called Halo CMEs, are the most significant because they slam the Earth's magnetosphere with violent shock waves. If solar storms are the source of MLs, then we should expect occurrences of Halo CMEs and MLs to be correlated. In Table 4, we show Halo CME events that occurred during a nine-day period prior to recorded ML events and find that the two types of events do not appear to be related.

Comment -- That MLs and Halo CMEs do not appear to be related is an obstacle to this hypothesis. However, the fact that the magnetosphere on the night side is uniquely different from its daytime state may still prove to be the contributing factor that

174

Table 4: Halo CMEs and ML events are not related. Dates shown are Universal Time (UT) and differ from Central Daylight dates by 5 or 6 hours depending on Daylight Savings Time, making UT dates the same or one day later.

Solar Halos Indicate Earth Directed Coronal Mass Ejections

9	8	7	6	5	4	3	2	1	Date of ML Night	# Halos in Prior 9 Days	# MLs Seen On This Night
							1	3	11/25/2000	4	1
						1	3	3	11/26/2000	7	1
					1	3	3	1	11/27/2000	8	1
	1					1			2/10/2001	2	1
				1					2/1/2002	1	1
									4/1/2002		1
			2	1	1				10/31/2002	4	1
			1						11/22/2002	1	1
									2/20/2003		1
									8/13/2003		1
									5/8/2003		1
									5/9/2003		1
									7/13/2003		1
									5/9/2004		1
	1								6/3/2005	1	8
									11/14/2005		1
									12/17/2005		1
1									7/15/2006	1	1
									7/16/2006		1
									7/17/2006		1
									7/30/2006		1
					1				8/11/2006	1	1
									10/20/2006		2
									1/1/2007		1
									3/14/2007		1
									6/20/2007		2
1									8/8/2007	1	1
									8/20/2007		1
									9/27/2007		2
									10/23/2007		1
									1/13/2008		2
									3/12/2008		1
									3/27/2008		6
2	2	0	3	2	3	5	7	7	Totals	31	49

makes MLs nocturnal.

Hypothesis 2: MLs are plasma descending out of the Inner Van Allen Belt.

Our very first satellite launched in 1958 discovered that the Earth is encircled by two belts of radiation known as the Inner and Outer Van Allen Belts (named for the scientist who discovered them). They are invisible; we look right through them when we look at the stars in the sky, but these belts contain high energy particles. The Outer Belt contains solar wind particles that have become trapped in the Earth's magnetosphere; it is these particles that result in generation of the Northern and Southern Lights.[6]

Inner Belt particles come from a different source, are created in a different way, and contain higher energy. Powerful cosmic rays* originate beyond our galaxy and periodically strike the Earth's upper atmosphere with sufficient energy to shatter molecular structures. The products of these collisions -- free ions, free electrons, and gamma rays** -- reside in the Inner Belt. Is the Inner Belt the energy source behind MLs? This concept was a primary speculation in my *Night Orbs* book for several reasons.

By May 2003, I knew for sure that MLs were capable of moving cross-country for miles. These cross-country journeys included periods when an ML would go out completely for a period of time and then return with a brilliant burst of light, followed by a resumption of its journey with renewed energy. ML activities on May 7th and 8th of 2003 were good examples (shown in Figures 34-38) as was the ML photographed on February 19,

*Cosmic rays are electromagnetic radiation of extremely high frequency and energy emitted by stars located beyond our Milky Way galaxy.
**Gamma rays are high-energy photons, especially as emitted by a nucleus in a transition between two energy levels.

2003 (shown in Figures 30-32). The way these traveling MLs were going out and then flashing brightly back into existence was consistent with the notion that they were being replenished during their off states, as they traveled cross-country. It was one thing to envision gas venting to the surface to ignite and sustain MLs in fixed locations, but not so easy to postulate moving MLs being fed from below as they skimmed along well above desert foliage. This second hypothesis might explain en route replenishment because the plasma would be descending from above.

The process of cosmic rays bombarding our upper atmosphere to add particles to the Inner Belt is continuously ongoing and the Earth's magnetic field can only hold so much; there has to be a relief mechanism. I speculated that "mirror points" (where spiraling trapped particles reverse direction) were getting pushed ever lower by overcrowding, until particles encountered the upper atmosphere, causing them to slow and spiral down to the Earth's surface.

Magnetic lines of flux all pass through the center of the Earth rather than through the North and South poles. That means that some of the inner dipoles*** do pass through lower latitudes. An excellent text book by Dr. Martin Walt, **Introduction to Geomagnetically Trapped Radiation,**[7] provided information and equations that made it possible to compute the top of the dipole that intersects the Earth at latitude 30 degrees (Mitchell Flat). That dipole fit well within the Inner Van Allen Belt providing a ready potential source of plasma.

This concept might explain why MLs occur during nighttime only, because unloading into the atmosphere should logically take

*** The term "dipoles" refers to magnetic lines of flux that are oppositely charged at two points, or poles.

place on the night side of the Earth, where elongation of the magnetosphere weakens magnetic forces.

Based on NASA satellite data, ions in the Inner Belt are estimated to have energies in the 10 to 100 million electron volts range.[8] Energy levels this high would account for observed energetic ML displays. Plasma displays might be capable of multiple transitions in and out of light-emitting states depending on concentrations and energy levels.

This budding theory required generation of an opposing magnetic field in Mitchell Flat to account for the fact that these descending particles did not simply disappear into the ground. Limited hand meter measurements on September 13, 2003, and reports from other researchers (including one report from Min Min, Australia[9]) did suggest that magnetic anomalies might be associated with ML appearances although to date I have not been able to document the presence of an opposing magnetic field.

Pro --

1) This hypothesis was able to account for ML's nighttime-only appearances.

2) The inner radiation belt is a very high energy source (clearly important given observed ML behaviors).

3) Continuous cosmic ray bombardment demanded that the inner radiation belt had to have a release mechanism.

4) It was a concept that might account for ML replenishment while moving cross-country.

It all amounted to an incomplete theory but one that was starting to make a certain amount of sense with potential to evolve into something more solid. That was my view in 2003 but what I have learned since then undermines this theory in a number of ways.

178

Con --

1) I now suspect that sprites (a high atmosphere component of some very powerful lightning discharges) may constitute a workable relief mechanism for inner radiation belt particles, jeopardizing the notion that trapped particles are being pushed into the atmosphere by overcrowding (space weather experts are still conflicted on this point).

2) Photographs taken with diffraction grating filters have shown at least one ML was plasma (see Figure 54), but others have shown continuous spectra (e.g., Figures 40-42). If MLs are plasma out of the inner radiation belt, then all spectra displays should be discontinuous. This item alone punches a significant hole in this particular theory.

3) Long ML durations (Story 10 = 7 hours and Story 26 = 3 hours) are inconsistent with this concept of particle unloading.

4) The enormous size and duration of MLs encountered on the night of June 2nd and 3rd, 2005 (Story 26), fed by the electric storm that night, seems inconsistent with the concept of plasma descending out of the Inner Belt.

Comment -- This hypothesis appears unlikely to be correct.

Hypothesis 3: MLs result from liberation of pyrophoric chemicals.

When two chemicals autoignite on contact, they are said to be hypergolic. Pyrophoric chemicals are a subset involving autoignition of a single chemical whenever it comes into contact with oxygen in the atmosphere. A pyrophoric substance that will ignite spontaneously as soon as it is exposed to air is a possible explanation of an ML's characteristic ability to turn on and off repeatedly. Pyrophoric chemicals can be vapors, liquids or solids.[10]

MLs photographed in February 2003 (Figures 30-32) as well as binocular observations (Story 28) have revealed the appearance of burning chemical processes. An obvious hypothesis is to assume that some suitable pyrophoric substance is embedded in the thick layer of tuff that covers Mitchell Flat and that, from time to time, pockets of these materials are liberated and forced to the surface and into the atmosphere where they may autoignite to create ML displays.

Why should this be a nighttime only occurrence? That may depend on the mechanism leading to release of the chemicals. Marfa is high desert; differences between daytime temperature highs and nighttime temperature lows can be significant. Visitors quickly learn that even though they may have sweltered during the day, they need jackets when they go hunting MLs. Perhaps these relatively large temperature shifts are necessary to the release of pyrophoric chemicals into the atmosphere (see Appendix B for ML and weather related data charts).

Another possibility is that warm gases might develop greater relative buoyancy at night when above ground atmospheric temperatures become significantly cooler than underground gas temperatures.

Pro --

1) This hypothesis offers an explanation for repeated restarts (autoignition of pyrophoric elements) and that is a solid contribution.

2) For reasons listed above, this could conceivably be a nighttime-only occurrence.

3) Other observed characteristics such as color changes, dancing antics, variable altitudes, varying intensity, and energy expenditure are all possible attributes, if we assume this type of

180

light source.

Con --

This hypothesis might work for Type II MLs because they are stationary events, but what about Type III MLs?

1) How do we explain cross-country travel into the wind?

2) How do we explain why such a light would not bob with wind gusts?

3) How do we account for energy requirements of long duration and long range travel including en route replenishment?

Comment -- This hypothesis does not stretch far enough to account for the full range of observed ML behaviors. However, obtaining a decent spectral profile of MLs may lend support to this theory if we identify constituent elements that are pyrophoric.

Hypothesis 4: MLs are electromagnetic vortexes that burn chemicals to produce light.

Mystery Lights differ in several important ways from a rare phenomenon called ball lightning, but the two may be close cousins. An offshoot of thunderstorm lightning, BL, unlike an ML, lasts only a few seconds, sometimes ends with an explosive discharge, and appears in daylight as well as nighttime.

It is clear that MLs are not ball lightning, but ML characteristics (Tables 1-3) do suggest a degree of commonality, especially with respect to dynamics of behavior (e.g., BL is reported to float horizontally, and show wind resistance).[11] Most importantly, aspects of the conceptual model used to explain BL may be useful to an explanation of Type CE MLs. A magnetic field is thought to surround and intertwine with the lightning offshoot that becomes BL. This fourth hypothesis assumes that the heart of every CE ML is an electromagnetic vortex (i.e., magnetically encapsulated electricity). These electromagnetic vortexes are not BL, and they do not, by themselves, emit light. Instead they provide an ignition source for flammable gases that may on occasion come into contact with them. Combustible chemicals entering these vortexes may be ignited anytime air/fuel ratios are satisfied. Chemicals burning within the vortex would then emit light and become MLs. Variability of fuel/air mixture could account for an ML's variable light intensity as well as starts and stops.

My ML time exposures have shown gaps in lights tracks at altitudes where obstructions were unavailable to account for these gaps. Figure 37 shows an ML flying at an altitude of ~250 feet, high above any available obstacles. What may have been moving through gaps in the light track was an electromagnetic-vortex with sufficient electrical content to ignite flammable gas whenever air/fuel ratios in its path were again satisfactory to support ignition.

These vortexes would be responsive to ground current magnetic fields and that may account for an ML's stability in wind currents and movement in directions not dictated by the wind.

What would be the origin of these electromagnetic vortexes? I suspect Earth (telluric) currents may be the source. There is a molten core located deep beneath the Earth's mantle that is spinning faster than the earth. This spinning core generates not only the Earth's magnetic field, but also induces electric currents that flow continuously through the ground beneath our feet.[12] Many natural sources contribute to these underground currents including electrochemical activities, atmospheric electricity, ionospheric conditions, and solar activity.[13]

Telluric current flows are complex because the Earth's crust has many different materials with widely varying electrical resistances. Variations in conductivity of Earth materials cause these telluric currents to be very concentrated at some locations throughout the world and there are reasons to believe that Mitchell Flat may be such a location.

Electrical environment of Mitchell Flat -- There are a number of reasons to suspect that Mitchell Flat has an unusual electrical environment. A case in point is the halo ML recorded by Roofus and Snoopy on June 2, 2005 (see Figures 43 and 44). That halo measured almost a mile across and half a mile high as captured from two different camera angles. Video analysis suggests that this halo was physical reality and not simply an optical illusion. As such, this halo may be visual evidence indicating the presence of a powerful magnetic field. That ML started with a lightning flash and seemed to be fed by subsequent strikes.

The ML that followed later that night was the largest orb that has been recorded by my research effort (see Figures 45 and 46). It was big and bright enough to light up overhead clouds and apparently impressive enough to attract the attention of military pilots (as previously discussed in Story 26, ML Behaviors chapter). All of the MLs that night occurred during a raging thunderstorm and appeared to be fed by the energy of that storm. MLs appeared and sometimes became stronger in connection with lightning strikes.

In the early morning hours of June 3, 2005, during one large lightning event, a separate lightning bolt was photographed crossing a small valley. This may have been ground to ground lightning (see Figure 83). As the storm moved out of the area, the large remaining ML lost energy and died. Videos from that night suggest that the prevailing electrical storm may have been an essential aspect of ML generation.

There have been a number of other indicators of electromagnetic involvement with Type CE MLs. On two occasions I was able

Figure 83. Possible ground to ground lightning bolt in Mitchell Flat (picture enchanced to better show bolt).

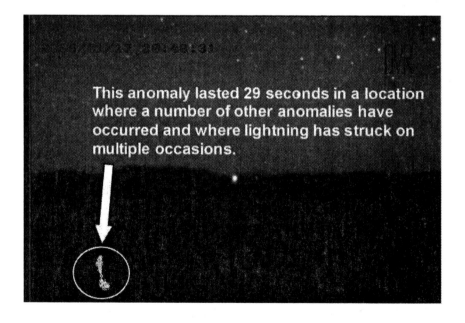

Figure 84. Suspected electrical anomalies have been recorded often in Mitchell Flat by night monitoring stations (picture enhanced to better show anomaly).

184

to detect magnetic disturbances associated with ML events using hand-held gauss meters (see "The Search Intensifies" chapter and Story 27). There have also been instances of electromagnetic anomalies photographed by night monitoring stations. The anomaly shown in Figure 84 is one example of a number of unexplained anomalies that have occurred in an area that has experienced repeated lightning strikes.

In addition to all of the above electrical instances, we must add electrical events involving vehicles and one train near Paisano Pass (see **"Another electric surprise"** in the **Night Discoveries** chapter). The vehicle-related anomalies have occurred where US 90 crosses above railroad tracks leading to Presidio. Those railroad tracks, which pass through the middle of Mitchell Flat, may have been the source of conditions that precipitated the anomalies.

To summarize, evidence points to existence of an unusual electrical environment in Mitchell Flat. This hypothesis is based on the notion that a natural concentration of telluric currents in Mitchell Flat has set the stage for generation of electromagnetic vortexes that may occur on occasions when Earth currents are surging. Lightning strikes add electrical energy directly into the ground and may produce surging of currents in regions were telluric current flows are most concentrated. Because telluric currents are global in nature, contributing electrical storms need not be in Mitchell Flat to produce current surges there.

Just how these electrical vortexes might be generated is not known. However, as the Earth spins, Mitchell Flat rotates from daylight into dark and the Earth's magnetic field elongates because the magnetosphere is significantly stretched on the night side of the planet. This rapid stretching of the magnetic field may play a role in "popping-out" electromagnetic vortexes and, if so, that might explain why MLs are nocturnal. It is also conceivable that intense telluric current flows are ionizing and liberating earth materials that rise into the air. If a rising cloud of ions becomes dense enough, then ground current flows seeking an electrical path with less resistance might leave the ground and flow through these airborne ions. If that happens, disturbing air currents could lead to creation of electromagnetic vortexes. That might be one reason why MLs seem to favor windy nights.

Chemical Considerations -- As stated in Hypothesis 3, I suspect that Type CE MLs involve chemical-combustion processes primarily because photographs (Figures 30-32, 37, and 53), as well as personal observations by multiple witnesses, have recorded and reported burning orbs of light with "combustion-like" attributes. But what is the source of flammable gas to support this concept? One possibility is hydrocarbons. Natural gas under the surface is a certainty: gas and deep oil deposits have been located by oil exploration companies, a few wells have been drilled, and oil well gas fires have burned from time to time. Mitchell Flat is a region with geological faults and those fault cracks may provide escape routes for hydrocarbons trapped deep beneath the surface. There may also be other sources of fuel in Mitchell Flat, given the uniqueness of its geological character.

I have searched for combustible gas sources using various gas meters (e.g., Figure C12, Appendix C) with limited results. Most gas readings have involved fairly low concentrations of gas, but on one occasion in 2005, my M40 meter detected and saved to memory a reading of 29% of the lower explosive limit (the meter was designed to measure explosive risk). Unfortunately, it was not clear exactly where the reading had been taken and it was not obtained again. A better meter and more precise data collection

186

strategy will be needed before we can know if combustible gases are periodically being vented in Mitchell Flat. For now, this continues to be a possibility.

Pro --

1) This hypothesis seems to best fit the entire range of Type CE characteristics, including nighttime only, variable intensity, turning on and off, wind resistance, splitting and merging, and dynamics of motion. Subtype I (tiny sparks of light too small and fast to photograph but with a remarkable capacity to travel into the wind) are seen during twilight and may even constitute events pointing to formation of electromagnetic vortexes as Mitchell Flat rotates into darkness and the Earth's magnetic field is stretched by the magnetosphere.

2) If electromagnetic vortexes are causing flammable gases to combust and release ions, then the door is open for MLs to be either plasma or chemical oxidation (fires) and this may account for captured ML spectra that have been both discontinuous (indicating plasma) and continuous (indicating conventional combustion).

Con --

I want to emphasize that this hypothesis is my speculation. There is photographic and observational evidence to support the idea that MLs involve chemical combustion and/or light emitting plasma, but so far no data to support gas venting to feed these displays. There is also evidence that supports presence of an unusual electrical environment in Mitchell Flat, but the concept of electromagnetic vortexes has not been established.

Summary of Light Sources in Mitchell Flat

Table 5 provides my estimate of ML appearances in Mitchell

Flat, based on the following assumptions:

1) People visit the View Park looking for mystery lights every night and at least some of them manage to find lights they consider to be mysterious.

2). On average, my monitoring stations are finding 5.25 ML nights per year (see Figure B1) and, for the purpose of this calculation, I will estimate that I am finding only half of the ML nights in any given year.

Assumption 1 yields 365 nights that result in people seeing lights they believe to be of mysterious origin.

Assumption 2: 5.25(my average) X 2 (assumes I am finding only half of all ML nights) = 10.5 nights per year estimated.

10.5ML nights/365nights = 3% actual ML nights with the other 97% being misidentification of explainable light sources.

Some may protest that 97% is too high for misidentification and/or complain that 3% is too low to account for the many convincing ML reports they have read and they may be right. My experience has been that "genuine" ML events can be very exciting. People who are convinced they have witnessed something truly unusual, may be more likely to retell their stories and to write them down. I believe the stories I have included in this book are all genuine ML experiences with the possible exception of Story 16 (a vertical departure that I now believe may have been a helicopter flight).

It is important to recognize that genuine MLs have multiple sources with Type M (mirages) and Type CE (chemical-electromagnetic) being the more common. Mirage MLs can be very mysterious (and scary when they seem to chase cars) but the physics of their production are understood and, in that sense, they

188

are explainable and are genuine ML light sources. Type CE MLs also exist and this variety exhibit characteristics that continue to defy conventional explanation. It is the Type CE ML that so richly deserves further study.

When it comes to dealing with the unknown, a degree of skepticism is perfectly understandable. Most of us tend to dismiss unusual claims as misperceptions. We may be impressed with the skill of a magician, but leave the theater knowing it was only an illusion, not actual magic. And when we see Mother Nature produce unexplained events, we may think "This is surely an illusion or a misperception" but only until we know better.

Table 5. Results of this eight-year study suggest that most reported Marfa Lights are due to explainable light sources. It is estimated that mysterious lights appear on only a small percentage (3%) of nights. These unexplained lights most probably have multiple, sources including some that are truly mysterious.

Categories	Sources of Marfa Lights	Est. Freq.
"MLs" from Explainable Light Sources	Vehicle headlights Ranch Lights Stars Iridium Satellites Low flying aircraft Sprites (a type of lightning) Lightning Fireworks Train Lights USAF radar blimp Lightning bugs Brush fires and trash fires Oil fires Hoax lights Lights associated with work activity at night	~ 97%

MLs from Unexplained and/or Mysterious Light Sources	Type M (Mirages) ~1 to 1.5%
	Type LC (Light Curtains) trace %
	
	Type CE (Chemical-Electromagnetic) ~1.5 to 2%
	   Subtype I – tiny specs of light
	   Subtype II – stationary
	   Subtype III – move x-country
	   Subtype IV – above horizon
		~ 3%
		
		Appendix A Stories belong In this category (with possible exception of Story 16)

Table 6. Summary of ML observations, related questions and speculations.

Observations	Questions	Speculations and Comments
MLs have multiple sources.	What are they?	Type M (Mirages) Type LC (Light Curtains) Type CE (Chemical – Electromagnetic)
MLs are nocturnal.	Why is this?	Type M – Temperature inversions are more likely at night (see text). Type LC – Cool temperatures may be necessary to form ice crystals and/or rapid drop in temperatures after dark may be necessary to release of chemicals. Type CE – Unexplained. (1) Earth's magnetosphere elongates approximately 6X on the night side, but so far no ML relationship has been found to the magnetosphere. (2) Temperatures drop quickly after sunset, but no relationships to ML appearance have been identified.
MLs are usually yellow-orange but can be other colors and sometimes turn bright red before going out.	What accounts for differences in ML colors and color changes?	ML colors may reflect energy states and or chemical content. It is suspected that changes to bright red result from energy decay.
MLs are silent.	Mirages would be silent but why would Type CE be silent?	Suspect that Type CE may emit detectable sound if observer were close enough to hear it. So far, there have been no reports of sound.
ML events do not relate to solar storms, lunar cycles, or weather patterns.	What triggers ML events?	Type M = Temperature inversions Type LC = Unknown Type CE = Unknown
Electrical anomalies, possible ground to ground lightning and "Weird Stuff" (see text) have been detected in Mitchell Flat.	Are these oddities related to Type CE MLs?	Unknown, but probably yes.

191

Table 6 (Continued). Summary of ML observations, related questions and speculations.

Observations	Questions	Speculations and Comments
MLs vary in intensity and turn on and off frequently.	Why does this happen?	Type M = Instability of atmospheric lens. Type CE = Probably caused by variations in fuel supply and/or electrical surges.
MLs go out completely and re-ignite frequently during typical events.	What is the ignition source?	Type CE = Probably electric ignition or else pyrophoric autoignition when fuel is available.
Stationary MLs do not bob in the wind and moving MLs do not follow wind directions.	How can any airborne light not be influenced by wind?	Type M = Mirages are fixed or moving lights located elsewhere so we should not expect wind response. Type CE = Suspect magnetic or electrical fields control.
ML light tracks show gaps that cannot be explained by obstructions in all cases.	What aspect of Type CE MLs moves through the gap?	Hypothesis 4 suggests electromagnetic vortexes.
MLs exhibit splits and mergers.	What causes Type CE MLs to split or merge?	Electromagnetic influence.
Some Type CE MLs are tiny and burn only once (subtype I), some are stationary (subtype II), some fly cross-country (subtype III) and some fly high in the sky (subtype IV)	What accounts for these differences?	Different energy states.
Type CE ML locations do not repeat except on sequential nights.	If Type CE MLs are ground based, why wouldn't locations repeat?	Suspect ML events exhaust essential elements at point of origin.

192

Appendix A

Mystery Light Stories

Appendix A
Mystery Light Stories

An excellent source of Marfa Lights Stories is a booklet titled *The Marfa Lights* by Dr. Judith M. Brueske. Other interesting Marfa Lights stories can be found in *Mysteries & Miracles of Texas* by Jack Kutz, *Ghost Stories of Texas* by Ed Syers, and *Tales of the Big Bend* by Elton Miles.

In this section I will add to that wealth of stories by retelling or quoting (in italics) what has been shared with me -- by a wide range of observers -- over the past eight years of my Marfa Lights investigation. Of course, I have my own tales to tell and those are included. Collectively, these stories provide readers with a feel for the breadth, depth and amazing complexities of Marfa Light experiences.

Story 1   **San Angelo Standard Times**
Date: February 1945
Source: Microfilm copy of the original is available at the Angelo State University Library, San Angelo, TX

This report was written at a time when the Army Airfield at Marfa was still active. Following are key points contained in the article. My investigation is at odds with some of their observations but this early report is remarkably comprehensive and complete:

- Mysterious lights that defy detection of source and location have been seen periodically southwest of Marfa *for more than 40 years* [J. Bunnell: The direction 'southwest' of Marfa is questionable given that Mitchell Flat is located east and southeast of Marfa. This may simply have been a typo].

– *Rains in April or May usually bring out the ghostly lights where they are visible at night from highways around Marfa until the fall dry spell brings them to an end.*

– *These elusive lights change color and shape and frequently disappear completely for a few minutes.*

– They *can be seen only at a distance and from the ground.*

– Search parties trying to reach the lights see them fade and disappear after a few miles even though people located on US 90 continue to see them as brilliant lights.

– Last year (1944) a plane from the Army Airfield conducted an aerial search for an ML [ML, short for mystery light, is my terminology; it was not used in the news paper article] but failed to see it even though it was plainly visible to those on the ground.

– *A guess fixes the mystery light about 50 to 75 miles southwest of Marfa. Some say it is across the river in Mexico.*

– MLs may be white, bright red, blue, and yellow or green ruling out speculation that these are reflected automobile lights.

– *The most generally accepted theory is that the light is a vapor rising from a phosphorous bed somewhere in the Big Bend.*

– *However, scientists at Sul Ross College in Alpine refuse to substantiate that claim. They frankly admit that they do not know the source.*

– *The light was supposedly first seen 40 years ago by a Marfa resident, Dr. Monroe Slack* [Earlier reports do exist. See Figure 2 for one example].

196

Story 2   **Geologist's Report**

Date: March 1973

Sources: (1) Paper titled, "The Enigma Lights of
               Marfa, an Unexplained Phenomenon"
               by Elwood Wright and Pat Kenney, dated March 1-
               June 16, 1973, Marfa Public Library
               (2) Dallas Morning News, Sunday, July 4, 1982.

Witnesses: Elwood Wright and Pat Kenney

Two geologists, Elwood Wright and Pat Kenney, while on assignment near Marfa spent some of their spare time looking for MLs in Mitchell Flat. Their experiences in March of 1973 convinced them beyond any doubt that Marfa Mystery Lights were real phenomena of the "Unforgettable" kind. Their report is summarized below.

March 14

At approximately 10:30 PM, while they were returning to Marfa, a light was observed approximately one mile south of US 90 and estimated to be about the size of a basketball. It varied in intensity, turned on and off, and at its brightest was about as bright as car lights on high beam.

March 15

Wright, Kenney and Wright's two sons observed that the horizon toward the west, south, and all points in between were *"lighting up"* at short intervals between 9 and 10 PM. The zone above the horizon up to about 2 degrees elevation would light up for one to three seconds even though there was no lightning, sky overcast, or any other apparent reason for the lighting effect.

March 19

Three lights were observed in the direction of the Chinati Mountains. One light began swinging in an arc while another did a complete loop and approached the rocking light. These extreme antics convinced the observers that they were not seeing automobile lights.

March 20

Wright and Kenney went looking for lights along a ranch road. They parked on high ground approximately one and one half miles south of US 90 and waited to see if any lights would appear. Three horses west of the road suddenly bolted and began behaving in a frightened manner. At that moment two Mystery Lights were observed moving from southwest to northeast very fast. Kenney had the impression that they were traveling 150 to 200 m.p.h. The

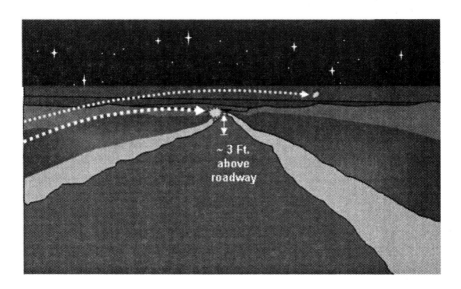

Figure A1. Mystery Lights seen by Elwood Wright and Pat Kenney crossing Nopal Road on March 20, 1973.

first light crossed the road approximately a thousand feet away and continued east where it seemed to meet and merge with a third light. The second light crossed the road close to where the first light had crossed, but the two geologists had moved closer by then and were only about 200 feet away when the ML paused and hovered, pulsing on and off approximately three feet above the road in front of them. Wright urged Kenney to *"floorboard"* their vehicle and try to run over it, but Kenney observed, *"Seeing how fast that thing could go, we can't run over it."* The intensity of the light decreased as it slowed down and hovered. The color was described as similar to an incandescent light bulb and the size was estimated to be about one-half the size of a basketball. After hovering for about 30 seconds the light resumed its journey east and joined the other two lights.

Comment:

This sighting is one of the best close encounter cases documented by professionals and both are credible witnesses. Their March 15 report of the sky "lighting up" has also been witnessed by the author and reported by other observers, but this could have been the result of a distant lightning storm located below the horizon.

Story 3 **Red MLs to the West**
Date: 1980's and later
Source: Fred Tenny - astronomer

Mr. Fred Tenny is an amateur astronomer with experience in night observations. His interest in astronomy started at an early age; he joined the Astronomy Club in junior high school and has for many years regularly participated in Star Parties (where amateurs and professionals alike gather to look at stars through high-grade powerful telescopes). He joined the Texas Astronomical Society of Dallas in 1999 and received an Honorary Messier Club Certificate at the May 2000 Texas Star Party. Today Mr. Tenny operates an active business supplying support equipment for amateur astronomers via his website www.gazeraids.com. Mr. Tenny's background in night observations lends credibility to his Marfa Lights experiences and comments. He has written the following:

Almost all the MLs I've seen are to the west of the present day View Park and appeared to be moving north. They have exhibited all the characteristics described in most accounts, bobbing up and down, splitting and changing colors with occasional retrograde motion back to the south. Most of them have been red, occasionally

Figure A2. Example ML located west of the View Park as photographed by Fred Tenny in the late 1980's.

fading to orange and very rarely yellow. I have seen individual ones split into as many as three, dance around each other and recombine.

I started going to see them in the mid-1980s and saw some every time I went. They seemed to be most active when there was storm activity in the area. Back then there was very little road traffic and the Marfa Lights View Park (MLVP) didn't exist yet. On at least two occasions they were bright enough and near enough to noticeably light up the ground near them. I did not visit the View Park during most of the 1990s.

Comment:

It is interesting that Fred has had best luck with seeing MLs located generally west (not southwest where cars on Highway 67 are visible) of the View Park. The second point of interest is his color description, "Most of them have been red, occasionally fading to orange and very rarely yellow." These colors contrast with white-amber-orange and occasionally red that I have observed to the south and southeast of the View Park. These differences intrigue. Locations west of the View Park would be flying over different terrain. If MLs come out of the ground and involve chemicals from the ground, then these color differences might result from chemical differences between the two physical locations. This is another clue to be considered.

Story 4   A Three-Ring Circus
Date: February 1991
Source: Paper titled "Seeing the Marfa Lights on February
              2, 1991" by Edson C. Hendricks, dated December
              27, 1991, Marfa Public Library

The following report by Mr. Edson Hendricks is one of the most detailed and spectacular accounts I have found. I have personally worked with Ed and can vouch for his veracity and his cautious, no-nonsense approach that is typical of most engineers and scientists (he is both). It is worth noting that when he went looking for Marfa Lights in 1991, he firmly believed that Marfa Lights stories were nothing more than misidentification of ordinary light sources. What he found was something much different. His report, written immediately after the event, provides an excellent example of complex ML behavior. It is summarized here.

It was a cold night on the second day of February when Edson Hendricks pulled into the Marfa Lights viewing area shortly before sunset. He planned to look for the lights with the aid of binoculars from inside his automobile to help protect against the cold and intermittent light rain.

As darkness gathered, Hendricks was able to identify stationary ranch lights in Mitchell Flat that were constant in brightness. Automobile lights on the Presidio to Marfa highway were also easily recognizable to the southwest. They traveled at relatively constant speeds northward winding through hills as they descended toward Marfa. The car lights varied in intensity and Hendricks correctly recognized that brightness increased when the roadway came into alignment with the View Park.

A few minutes before 7 PM, Hendricks noticed a bright light to the west-southwest, north of where he had been observing automobile traffic. At first this light remained still. It then began to move slowly north. The light was brighter than the automobile high beam lights to the southwest and it also flickered in an irregular way. The light gained speed, passed behind a utility pole as it moved north, and then paused for a minute or two. The light then slowly began to rise and reversed its course, moving back past the utility pole as it proceeded southward. At this point, Hendricks was puzzled by the light but still suspected it might prove to be an automobile. To his surprise, the light suddenly divided into two lights that continued southward while spreading apart. The time was 7:08 PM.

More surprises were in store. As the lights continued southward the left light increased in brightness and divided again while the right-hand light changed direction and began to accelerate as it moved northward, achieving a speed greater than the automobile traffic on Highway 67. This northward-bound light gradually faded and vanished.

Looking back to the south, Hendricks noted that yet another light had appeared and all were displaying erratic movements and variable brightness. This strange behavior continued until there were five lights. During an hour of observation these inexplicable displays became more complex and, in his words, "resembled a three-ring circus," as there were more concurrent activities than one person could follow. Hendricks observed at least six instances of single lights dividing, moving apart, and then subsequently merging back together. In one case, as two divided lights came back together, one light spiraled upward and seemed to circle the other light.

Hendricks also observed several instances of a light vanishing at the same exact moment as a new light appeared some distance away. He could not tell if the same light had "jumped" to a new location at terrific speed or if they were independent lights.

Hendricks also observed a deep red, dim light approach two brilliant yellow lights. As it drew close to one of the lights it flickered and flared into brightness, changing from dim red to bright yellow.

Sometimes the lights remained still, but mostly they moved both vertically and horizontally with the vertical movements being small compared to the horizontal movements. Hendricks could detect no directional preference in these erratic movements. For the most part, the lights seemed to move independently but sometimes they would form a line and move together. At one point, four bright lights disappeared together as if connected to a common off switch. They then came back on independently.

One pattern that Hendricks did notice was that lights would flare to brightness immediately prior to splitting. Bright cycles did not always result in splits but splits seemed to follow the bright phase. Another interesting observation was that the central point of the display of multiple lights seemed to be quite stable although it drifted to the south.

About halfway through this activity the lights started to fade and by 7:35 PM they were no longer visible. Hendricks was temporarily distracted by the arrival of other people at the viewing area. Getting out of his car to converse with the newly arrived people, Hendricks discovered that even more lights were then visible to the southwest. Car lights were also visible in that direction but could be distinguished by differences in color, brightness and pulse rates. All observers agreed that they were seeing the same ML displays. By 8 PM the MLs had once again faded and did not reappear, although ranch lights and automobile lights on Highway 67 remained in view.

Hendricks commented in his report that mirage effects might be a possible source but he considered that explanation unlikely because of color and behavioral differences with respect to the automobile lights observed on Highway 67.

Comment:

Instead of finding hoaxes or misidentification of ordinary lights, Hendricks was treated to a spectacular display that removed for him all doubt about the reality of Marfa's Mystery Lights and ignited his own research effort into these strange phenomena.

Story 5   Stormy Weather MLs
Date: 1990's and later
Source: Personal conversation in Alpine with Van

Van has lived in Alpine all of his life and has seen Marfa Mystery Lights many times. He remarked that his best sighting occurred one year in September when he had gone to Marfa to attend the annual Marfa Lights Festival. On returning to Alpine after dark he noticed that many cars had pulled off the road. People were out of their cars, facing south, and watching light displays and thunderstorms in Mitchell Flat. He stopped and joined them to observe a magnificent display of MLs dancing, splitting, merging, orbiting, spinning and doing all sorts of antics in the midst of the thunderstorm that was producing rain and lightning. Van added that the best displays were always on cloudy or stormy nights and he could not recall ever seeing MLs on a clear night.

Comment:

Van's comment that stormy weather is required for ML's echoes the comments of a number of other ML witnesses even though some take the opposite view and maintain that clear nights are required.

Based on my personal observations, MLs are seen on both cloudy/stormy nights and clear nights but it may be possible that weather is a factor. Additional data collection will be necessary to come to any conclusions regarding weather and electrical storms.

Story 6   **The Man Who Touched an ML**
Date: 1994, winter
Source: A personal story told to me by Minister Alton Sutter and to a video production company, Actuality Productions
Witnesses: Two ministers, their spouses and many children

In the summer of 2003, I participated in the filming of one segment of a Discovery Channel program titled "Miracle Hunters." The program involved different kinds of unusual happenings and one seven-minute segment was devoted to the Marfa Lights. The film crew recorded eye witness stories of ML sightings, including an exciting story by a minister, Alton Sutter. I was not present for that part of the filming but met the Reverend Sutter and the two involved families at dinner. It was there that Mr. Sutter told me about their most unusual ML encounter.

The two families had driven out onto a ranch road at night in hopes of seeing Marfa Lights. They were not disappointed. A group of lights flew over the roadway near their location. Most of the lights continued on course but one light slowed, causing Sutter's son to announce, *"Look Dad! One of them is landing."* That was the case, as one light began to shrink in size as it descended onto the far side of the roadway. Mr. Sutter was the first to reach it and by then it had reduced to the size of a pea. I asked Sutter how large it had been while still in flight and he said it had initially been about the size of an orange or softball. I asked Mr. Sutter if the light ball had been solid, transparent or translucent. After reflection he said it was translucent. He went on to explain that he took off his glove and scooped up the small remaining sphere and held it on the tip of his finger.

All of the two family members gathered around to look at this strange remnant of the light. As they watched, it continued to dissipate until it vanished completely, leaving no residue. I asked Mr. Sutter if it had been hot to the touch, but he had no recollection of heat. He did note that it was a very cold night and his hand was "freezing cold."

Comment:

As Minister Sutter told his story, members of the two families were all nodding their heads in agreement. The fact that there were so many firsthand witnesses, that they were all present when the story was told, and that they all indicated agreement with the story makes this particular account unusually well supported. Unlike lights in the sky that are subject to misinterpretation, this encounter involved physical touch in the presence of many witnesses. I do not know where we could ever hope to find a better "first-hand" account than this one.

The story is profoundly important on multiple levels. The lights that flew over the roadway were extremely close and one became close enough to touch. This flies in the face of skeptics who argue that Marfa's mystery lights are only misinterpretation of automobile lights. Mr. Sutter was not holding an automobile on the tip of his finger. The fact that there was physical contact also dismisses insistent voices of a few who maintain that Marfa Mystery Lights are only night mirages and/or tricks of refracted light. I believe it is true that night mirages are the source of many ML reports, but this account provides confirmation that there is more to this mystery. From a scientific point of view, the existence of an ML remnant that could be picked up on a bare finger while it continued to evaporate, leaving no residue, is important in its difference from other accounts, but also it raises more questions regarding the nature of ML phenomena.

Story 7   **Undulating Plasma Balls**
Date: Late summer 1996
Source: Email story submitted to editor@nightorbs.net
               [Edited for length]
Witnesses: Dirk & Sarah Vander Zee

My wife and I saw them and we had no idea they are as rare as has been reported. Apparently, we were there on a "super outbreak" night. Filmed them on my Sony camcorder. Highly doubt they were mirages. Behavior was best described as "undulating plasma balls" that folded in on each other, then split, blossomed, divided, rejoined, re-split, then dimmed, bloomed again, then blinked off. Like spherical torches being suddenly snuffed out. It was a totally moonless night.

Many repeats [of these cycles], separated sometimes by 10-15 minutes. Viewed them for 2 plus hours, from both the "viewing area" and also down south along Nopal Road. Totally bizarre. Certainly not the Presidio highway lights which were easily seen off in a different direction. These were of a much different spectra [color], and more importantly, visible through the telescopic view finder, and in a totally black field of view. We could actually lock on and focus [the camera] on them on most occasions, and they were [performing] some sort of internal movement, best described as multi-spectrum light folding in on itself, spherically shaped.

On return to the ranch house where we were staying that night we saw several depressed areas in the ground that glowed a soft, ethereal blue light, as if some dim blue gas were poured into these small depressions about the size of a swimming pool, and left to slowly be absorbed into the ground. We were unable to film these [blue] events, as they were below [the minimum] light threshold of the camera.

No explanation. Certainly not headlights. No way a mirage, as we were literally within feet of these particular phenomena.

Theory: I am no Earth scientist, nor geologist, nor spectral-analysis guy. Just an individual with no particular dog in this fight. I think the lights are electrical in nature, based on some as-yet unexplained surface Earth plasma current. I don't think they are "gasses" although the pools of blue light were stunning. It is some form of electrical or plasma energy "field" that has a surface exit point near Marfa.

Or not. Go see for yourself.

Comment:

Dirk provides a clear description of the Marfa Lights phenomenon consistent with my own observations. What makes this report special is reference to the pools of blue light existing in ground depressions later that night. Also, his observation, *"...some sort of internal movement"* within the lights is interesting. Both are keen observations of important details.

Story 8   MLs Southwest of View Park
Source: Telephone conversation with rancher M

A long-time Marfa resident who has ranched in Mitchell Flat all of his life commented that he has seen MLs on many occasions. He reported that he has always sighted them in the general vicinity of Antelope Springs, located southwest of the View Park in Mitchell Flat. Asked if he meant north or south of the Presidio railroad tracks, he replied *"both."*

Comment:

The region of Mitchell Flat that the rancher refers to has been a common location for people who have had reported close encounters, including Alton Sutter. The best view of this region would be from the Marfa Lights View Park and that is why the View Park is located where it is. Unfortunately, this region of interest as seen from the View Park does align to some degree with the region where automobile lights can be seen on Highway 67. That is unfortunate because looking in the direction of Highway 67 is confusing at best. Even with my experience, I avoid looking in that direction unless I am tracking lights that originated too far left or too far right to have been from highway traffic. MLs tend to be more orange in color and to do more pulsing than vehicle lights but my recommendation is always to locate Highway 67 lights and avoid looking in that direction.

Story 9	**"We Saw the Marfa Lights!"**
Date:	November 2000
Sources:	(1) Paper titled "We Saw the MARFA Lights!"
	Sharon Eby-Martin, Marfa Public Library
	(2) Email exchanges with Sharon Eby Cornet

November 24

On Friday night, the day after Thanksgiving, 2000, Sharon and her family arrived at the View Park east of Marfa shortly before dusk. As it grew dark, mercury vapor (MV) lights located near ranch houses in Mitchell Flat switched on and were easily recognized. Automobiles streaming north from Presidio to Marfa along Highway 67 were also clearly visible to the southwest. Sharon noted that the View Park was "loaded" with visitors hoping to catch a glimpse of the fabled Marfa Lights.

About 6:45 PM, Sharon's nephew Derek saw a very bright light "turn on" south-southwest of the View Park. At first they thought the light was moving, but Sharon was able to line up the light and a nearby utility pole to see that the light was, in fact, stationary. The light was too far east to be an automobile light on Highway 67. As they watched with fascination, the unknown light changed from yellow to orange to red and occasionally to white. They noted that it would also "blink out" and then later turn back on.

After making a trip to town, Sharon and family would be surprised to find the View Park completely packed with people. They decided to venture down a ranch road into Mitchell Flat itself in hopes of finding an uncongested view point. Being careful not to leave the county-maintained road (ranchers in West Texas are quite serious when it comes to preserving personal property rights), Sharon and her family found what looked to be a good observation point. They parked and observed from inside their warm SUV. From this vantage point, they saw two stationary lights. Getting out of the vehicle to get a better look, they were surprised to see that the two lights had doubled: there were four lights! Distance to lights at night in the desert can be very deceptive but Sharon and

211

family members believed those four lights were probably no more than a mile away.

Watching through binoculars, they were amazed to see the four lights merge into two and then split again into three lights only to merge again into two. Clearly, they were not seeing ranch lights or automobile lights. In fact, Highway 67 was located to the far right of the direction they were looking to see MLs.

Running the engine of their SUV to keep warm was consuming gasoline and Sharon returned to Marfa for more fuel. After refueling, they returned to Mitchell Flat. As they drove the hilly ranch road, they could once again see the two mysterious amber lights but, because of terrain obstructions, the lights appeared intermittently at first, and then disappeared entirely. Sharon drove as far as the road would permit without seeing the lights again. Turning her vehicle around, she was surprised to see the two lights were again visible. How did that happen? They continued to try to drive closer to the lights but, like chasing a rainbow, could never seem to get there. First one light faded and disappeared and then the other.

They gave up their ML observations at about 12:45 PM.

November 25

On the next night, Sharon, her oldest son, and nephew resumed the search. As the desert grew dark, the two amber-colored MLs once again made an appearance. Watching through binoculars, they could see the distant amber lights standing stationary and repeatedly pulsing from dim to very bright to dim. In the bright phase, they would light up surrounding brush and terrain for a radius of at least a hundred feet. By the time they headed home, Sharon knew, at least to her satisfaction, that Marfa's fabled Mystery Lights really do exist and are, as advertised, mysterious.

Comment:

On November 25th, wife Marlene and I were at the View Park watching the same pair of lights that Sharon and family were seeing from a ranch road, as will be related in the next story. We were unaware of Sharon's presence and we had not yet met.

Story 10   **Getting Hooked**
Date: November 2000
Source: Personal account: Jim Bunnell and wife Marlene
               (now deceased)

November 25:

My wife and I were making monthly trips to Carlsbad, New Mexico, to check on my father. In November 2000 we traveled to Carlsbad for a Thanksgiving visit. After Thanksgiving we decided to take a side excursion to Marfa to look at the Mystery Lights. I had not seen them in over 40 years and my wife, Marlene, had never seen them. She was curious, so we traveled to Marfa and arrived at the primary viewing site at sundown. The old Army airfield where I played as a child was long gone. We parked nearby at the View Park and watched the rapidly gathering darkness for any unusual lights.

With no clouds and no moon on that Saturday night, there was a dark clear sky, ideal for viewing lights. As it became dark we could see fixed red lights (both flashing and steady) and fixed green lights. These lights were constant in their behavior and were clearly not mysterious. We could also observe distant car lights that would pop in and out of view as vehicles traveled along winding roads through hilly terrain lying mostly southwest of our viewing location. These automobile lights were not difficult to distinguish because they followed repeating patterns of motion and moved only to the right as they traveled Highway 67 on their way from Presidio to Marfa. None of these lights could be considered mysterious.

Soon after dark we saw two strange lights on a compass bearing almost due south from our viewing location. These lights pulsed independently and seemed to follow a random sequence that, in most cases, went from dark to relatively dim, flared to a higher level of brightness, then dimmed and eventually went out. Sometimes both lights would be on at the same time. The lights were orange-white although the one on the left did turn orange-red

Computer simulation of Mystery Lights seen by Marlene and James Bunnell from their car on the night of 25 November 2000.

Figure A3. We saw two stationary mystery lights that turned on and off and had variable intensity (computer simulated as we had no cameras at that time).

for about two cycles. Both lights were well below the horizon and they appeared to be quite close to our viewing site. The one on the right was somewhat lower than the other and frequently would dim so much as to become invisible to the naked eye even though it could still be detected through seven-power optics as a glow on surrounding brush. The lights may have had minor horizontal and/or vertical movement but that could not be determined with our limited equipment. Sometimes they did appear to descend behind an obstruction such that only an indirect glow could be seen (or else an obstruction was raised in front of them creating the same effect). This appearance of vertical motion was more frequently observed with the lower light (the one on the right). Figure A3 is a computer simulation of what we were seeing.

We were not alone at the viewing site. A number of other vehicles came, left, or stayed and people milled about and commented on the same lights we were observing. I took another bearing on these same lights from one mile east of the primary site to accomplish triangulation. We continued to view these two lights for about two and one-half hours and then returned to our motel in Alpine. I returned to the site at about 1:00 AM that same night and resumed observation of these two lights until they extinguished at 2:30 AM. Although I remained at the viewing site until approximately 4:00 AM., they did not reappear.

From my youth I knew Marfa's Mystery Lights are truly mysterious but after our previous night's viewing both Marlene and I could not help but wonder if Marfa citizens had arranged some sort of light display to encourage tourism. The lights we observed Saturday night were few in number, apparently located very close to the viewing site, came on promptly after dark, went off completely at 2:30 AM., and exhibited only small movements, if any.

215

This pattern was too consistent and a little too pat. Were these lights being artificially produced to attract viewers?

November 26

Sunday night I returned alone to the viewing site determined to find out if these strange displays were maybe not so strange after all. I arrived at the primary viewing site shortly after dark and found it once again covered up with other visitors already intently watching the night's light displays. Mysterious lights were again visible from the same location and continued to play their "now you see me, now you don't" pattern we had observed the night before. To my astonishment, on this second evening they were accompanied by numerous other lights located to the left and the right of the original location. In total, there may have been as many as eleven lights with as many as eight emitting simultaneously. This array of lights ranged from about 168 to 258 degrees magnetic from our location at the View Park. In the spot occupied the previous night by the single right-hand light, there were now two and sometimes as many as three bright lights arranged in a closely spaced horizontal row. The total number of lights observed on this second night, and their wide distribution, cast considerable doubt on my suspicion that they might be artificially generated.

I was very tempted to walk to these lights, but I recognized such an attempt would be foolish given the dark night (again no moon) and the certainty that lying between my location and the target area were many cacti, barbed wire, and possibly snakes, not to mention cow patties. It was unpleasant to even think about walking into a cactus wearing Nike running shoes. Also my clothing was not adequate to withstand the Marfa night air with temperatures in the low 20s and made colder by wind. Moreover, I had

216

no permission to enter (trespassing is illegal) this private land.

Comment:

In Story 9, Sharon Eby reported that she and her family were never able to reach the lights because they would go out when approached. That would be characteristic of a mirage, except for the fact that on November 25th Marlene and I observed these same lights from the View Park. Sharon's view point and mine were separated by at least five miles and our viewing angles may have been separated by as much as 80 degrees; the mirage explanation does not work here. See Part II for more discussion of mirage MLs and non-mirage MLs.

Story 11   MLs and Small Red Lights
Date: February 9, 2001
Source: Email from Linda Lorenzetti [edited for length]
Witnesses: R.J. Creasy (deceased), Edson Hendricks, & Linda
                 Lorenzetti

Linda: *Looking through the www.nightorbs.net website again for new things, I came across one photograph that looks very similar to [the] MLs [that] Bob, Edson and I saw in February 2001. Since we saw them from the Marfa Lights View Park we may have had a different view but essentially the MLs we saw looked the same even including the two little red lights located right of the main ones.* [J. Bunnell: Figure A4 is the photograph Linda refers to in her account. The little red lights are above and right of the orange colored lights but are so small that they are difficult to see in this reproduction of my photograph. My photograph was taken from the MLVP on August 12, 2003.]

217

Our sighting occurred on February 9, 2001, around 7:30 PM CST. One of the reasons we'd often gone to Alpine in February was that Bob thought the conditions that time of year (cold, clear, and windy) might be conducive to seeing light phenomena. That night was the clearest that I can remember (don't believe I've seen it as clear since). Earlier in the day as Bob and I drove in from El Paso, the weather was very cold, windy and clear. Bob wanted to go out early in the evening to get situated before dark and as I recall we were at the viewing site around 5:30-6 PM. My field notes say that even the car lights on Highway 67 looked brighter (ceiling and visibility unlimited). I was sitting in the car and Bob and Ed were outside. I heard them talking about some unique light objects in the distance. By the time they called me to come take a look I could see four yellowish (yellow-red rather than light-yellow) lights, all on the same plane. These I saw with my naked eyes. Once out of the car, I viewed them through binoculars and equal spacing between the lights was pretty obvious. The lights began to wink out until only one light was visible and then it also disappeared leaving the area where the lights had been seen totally dark. My notes also say that the area was bounded, in a way, by ranch lights-one to the lower left of the target area and two to the right of the target area.

A short time later, (maybe 10 minutes) several of the lights reappeared, though I don't remember seeing all four of them appearing in exactly the same way. They were spaced farther apart and I cannot remember noticing the entire set of four visible together for as long a period. Most of the time only one or two were displayed. One light in particular became EXTREMELY bright with intensity that went from dim to almost blinding and then it disappeared. I don't recall seeing the two red lights on the

218

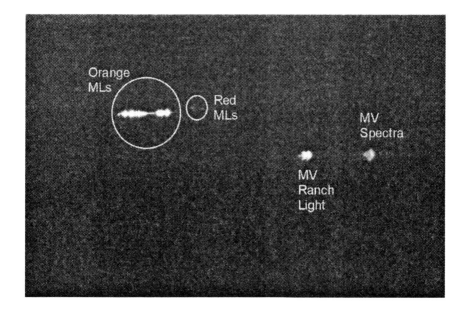

Figure A4. MLs photographed by the author on August 12, 2003 from the View Park. Two small red lights are located right of the orange colored (upper) lights, but are difficult to see in this reproduction.

right except toward the end of the displays. I don't remember the lights as being very far above the ranch lights, but I don't have a good sense of that. I was borrowing binoculars from Ed so there were times when Bob and Edson were using optical aids and I was not. We left the viewing site fairly early that evening, no later than 9 PM.

My notes also show that this was only the fourth time I'd been out looking for lights and that I'd never seen anything like these. Their color reminded me of sodium vapor lights.

Comment:

The similarity between Linda's February 2001 sighting and mine in August 2003 is remarkable with respect to location and especially with regard to the two tiny red lights observed above *and* right of the main body of MLs. Whenever I see repeating patterns I am always suspicious that they may be manmade lights because MLs have a wide degree of diversity in terms of location and configuration. Linda and her companions were careful observers, not likely to be fooled by artificial lights. The lights I observed in 2003 were persistent. They would turn off completely and then resume their antics an hour or two later. I took a number of photographs that night and was convinced at the time that they were MLs and not vehicle lights. Notice that the two red dots are well separated from the main body of MLs that were moving left. The small red lights did not follow. However, without greater optical magnification than was available at the time, it is not possible to rule out vehicle lights entirely so I now refer to these as ULs (unknown lights).

Story 12   High Flying MLs
Date: January 31, 2002
Witnesses: Bill, wife and granddaughter

Bill and his wife were staying in Ft. Davis at the Indian Lodge and had not planned a trip to Marfa to look at the lights but they purchased one of my earlier books, *Seeing Marfa Lights*, and decided to go take a look for themselves. They arrived at the park around 7:30 PM and parked near the eastern end. After observing for 10 or 15 minutes they noticed a light about 15 degrees above the skyline and a little left of center between two Mercury Vapor (MV) lights. It would climb in a scooping motion, and then come down as if it were rolling down stairs. It moved very suddenly from left to right and then back again, all the while becoming brighter and dimmer and changing colors from yellow-white to red. Vertical movements were estimated to be 10 to 15 degrees and horizontal movements were on the order of 20 to 25 degrees in a back-and-forth motion. The light was still visible when they left the View Park and other people were watching it.

Comment:

This sighting is important because it definitely involves aerial antics well above the horizon and was first sighted above the horizon. The altitude would seem to rule out automobile lights, ranch lights, and probably night mirages. The swooping, climbing, descending and rapid back and forth movements would also seem to rule out aircraft or starlight. Nevertheless, I did check star charts for the direction, elevation and time of this sighting. There were no potential candidate stars anywhere close to that section of the sky during the time of the sighting.

Subsequent to receiving this report, I learned of at least two other similar reports that definitely involved lights starting well above the horizon, doing loops around each other and then diving down into Mitchell Flat where they split apart and faded away. Because of their granddaughter, the observers elected to leave while the aerial display was still in progress. Had they stayed, it is possible they might have seen the light dive into Mitchell Flat and/ or split into multiple lights.

Story 13   No Wind Response
Date: March 30/31, 2002
Source: Personal account
Witnesses: James Bunnell and a number of other unidentified
              visitors at the View Park

221

On March 30, I traveled to Marfa, not to look for Mystery Lights, but to see the new View Park expansion and to obtain daylight range data of known objects in Mitchell Flat using a Swiss Army range finder. I arrived at the View Park at 11:30 PM, saw the freshly opened facility and talked to a few park visitors, but I saw nothing unusual.

The following day I used the range finder in daylight to more accurately locate regular light sources and various structures in Mitchell Flat. That night I observed a single fairly dim light about one-half hour after dark. Given its dim glow and direction, I suspected it was most likely an automobile on Nopal Road. This light stayed on continuously and, on a few occasions, briefly flared to high brightness. I continued to suspect this was only an automobile light. At one point a second light appeared to the left of the first light. The two lights were separated by perhaps 200 feet depending on how far away they were located. The two lights started to flash brightly together and then executed a merge into one light followed by separation again into two lights.

Following the interesting merge activity I began observing these lights carefully. The second (left) light went out but the original light continued to burn dimly and to occasionally flash to high brightness. The second light made a few more appearances but the merge event was not repeated. The original light continued to emit dim light with occasional flashes to brightness for at least two more hours. During this time I observed it with 20X tripod-mounted binoculars. It remained anchored in the same location even though a noticeable wind was blowing. Wind velocities were later determined to have been 8-12 miles per hour with occasional gusts to 22 m.p.h. I was impressed by the fact that the ML(s) never showed any reaction or response to the wind at all and remained rock steady with no indication of responses to wind gusts.

Unlike the left light, the one on the right never went completely out during the two-to three-hour event. The Swiss range finder could not be used in the dark and, without range information or a compass, I was unable to determine where the lights were located. Based on their unusual behavior, I became convinced that the lights were definitely not automobile lights. Times between flashes to brightness were of relatively long duration, much too long for the patience of someone wishing to hoax. I drove down Nopal Road and verified that the light(s) were not automobiles or ranch lights. Conclusion: These were genuine Mystery Lights.

Comment:

This sighting was perhaps the third time that I had noticed that multiple stationary MLs tend to string west to east with the more easterly position demonstrating less persistence and sometimes less intensity than the more westerly position. It also served to reinforce the observation that splitting of the lights always seems to occur during a bright phase while merging can occur during a dim phase. Most of all, this sighting was a clear demonstration of how MLs can hold station even in a relatively strong wind -- not only hold station but do so without any evidence of wind influence. The November 2000 sightings had also occurred in relatively strong winds with no observed wind response.

This sighting was also a reminder that it is difficult to validate ML sightings of short duration whenever time is not sufficient to establish strangeness. Had this sighting been of shorter duration, I would likely have written it off as a probable automobile simply because the location was near Nopal Road.

Story 14   Orange Light above the Horizon
Date: October 30, 2002
Source: Personal account

Witnesses: James Bunnell and two other View Park visitors
    I arrived at the View Park before sunset at 5:20 PM CST. The
temperature was 48 degrees F. with a chill wind blowing at 16.5
m.p.h. and gusting to 22.5 m.p.h. As sunset approached, wind
speed seemed to increase to about 20 m.p.h. and then decayed
some after sunset. To help ward off the cold I did my observing
from inside the View Shelter where a number of other people were
also braced against the wind. They were all watching vehicle lights
on Highway 67 to the southwest.
    After listening to a couple of obviously well-educated men
speculate on whether or not they were viewing Mystery Lights, I
joined them and pointed out that the repetitive nature of the light
displays was caused by automobiles negotiating a mountain road.
Because of automobile traffic to the southwest I do not normally
waste time looking in that direction but we were all looking to the
southwest as I explained the existence of and confusion caused by
Highway 67 traffic. The time was 8:55 PM when I noticed what
appeared to be an aircraft flying just below the horizon on a northwest trajectory. My two companions also saw this single orange
light traveling rapidly left to right at a constant altitude. The light
did not strobe as it moved silently and quickly through an angular
displacement of approximately 10 to 15 degrees and then blinked
out. The light could not have been an automobile because it moved
too rapidly and in a straight line with the Chinati mountain range as
a backdrop. The apparent speed of movement might have been
appropriate for an aircraft but as we watched, the light vanished

224

completely with no hint of aircraft navigation lights or strobe lights.

My two companions confirmed that they had seen the same thing I saw and speculated that it might have been a drug plane from Mexico coming in low to stay below radar. This idea was at least conceivable, but seemed unlikely for at least three reasons:

(1) The flight path was almost directly toward the Air Force radar balloon used to detect and interdict such traffic. Drug smugglers are well aware of this aerial radar and the direction that the aircraft (if it was an aircraft) was coming from has no clear passage through the mountains from Mexico.

(2) If it was a drug aircraft why would they be showing any lights at all?

(3) Being an aerospace engineer and lifelong aviation fan, I have observed aircraft flying at night countless times and the orange light we saw was uncharacteristic of any aircraft I have seen.

Comment:

This ML was definitely of the aerial variety and short-lived (only a few seconds). Without knowing range, it is not possible to know velocity, but unless the light was deceptively close to us, it must have been moving at 'aircraft-like' speed. The level trajectory and relative speed eliminate the possibility of a meteorite or ground vehicle. It is possible, but unlikely, that it was an aircraft.

Story 15   **Pulsing Light On Side of a Mesa**
Date: November 21, 2002
Source: Personal account
Witnesses: James Bunnell and four View Park visitors

This observation was made from inside the View Shelter and it occurred at approximately 7 PM CST before moonrise. There were a number of other witnesses present and we all agreed regarding what we were seeing. Background for this ML was Mitchell Mesa that extends approximately 500 feet above the basin floor and is located approximately 19 miles southeast of the View Park. The ML at first appeared to be positioned about halfway up the mesa. It was yellow in color and "pulsed" in typical fashion. The light initially remained stationary on a magnetic bearing of 157 degrees from our location at the southeast corner of the View Shelter. After perhaps 8 or 10 minutes of variable intensity flashing, the ML began slowly descending. It then extinguished and reappeared as a bright red light that lasted only 2-3 seconds. The light did not reappear after that.

Comment:

An extensive network of ranch roads and trails seem to go almost everywhere in Mitchell Flat but none could be found near where the ML was observed. The search for possible roads was extensive using both topographical maps and satellite photographs. Even allowing for significant error in the magnetic bearing measurement, no roads could be found that would work. The pulsing behavior of this light combined with unavailability of roads on the side of a very steep mountain slope makes it unlikely that this light could have been a vehicle. The sudden switch from yellow-orange to bright red followed by the light going out is also behavior that has been observed with other MLs.

Story 16   An Amazing Vertical Departure
Date: March 29, 2003
Source: ML Monitoring Station 1 (Roofus)

An unidentified light (UL) initially held position and pulsed for 27 minutes and 11 seconds. At 8:51 PM CST, it divided into two ULs and one of them rose into the sky and departed the area on a northeast heading. Trajectory of the flyaway was rocket-like in that it first rose vertically and then curved over into a fairly flat trajectory accelerating ever faster as it flew over a mercury vapor ranch light and on out of sight. The UL seemed to disappear as a result of distance from the camera instead of reaching the edge of the screen. A second (or third) UL remained weakly pulsing in the original location to the right of the mercury vapor ranch light.

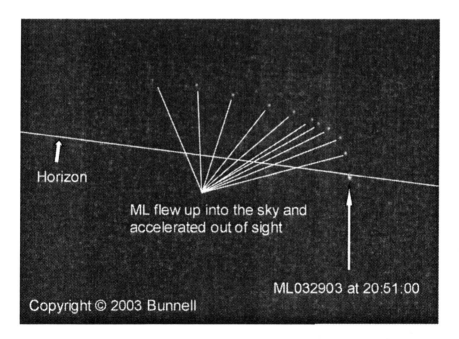

Figure A5. One light split from the primary and rose into the sky where it moved off to the northeast.

227

Comment:

This sighting by one of my nighttime surveillance cameras is unique because the UL's offspring rose up into the sky and flew away and there have been no similar UL sightings since.

This story was initially published in my book, *Night Orbs*, as an unusual ML. However, further analysis of this particular event showed the location of these unknown lights to be at the headquarters of a large ranch in Mitchell Flat. Suspecting this could have been a helicopter night flight from the ranch to Alpine, I contacted one of the ranch owners and asked him if he could recall such a flight. He said they do not normally fly out of the ranch at night and suggested that perhaps my camera had photographed a Border Patrol helicopter. That may be what it was. The actual source of this unusual event remains unknown but a nighttime helicopter flight to Alpine is a possible explanation.

Story 17   **Spectacular MLs**
Date: May 7, 2003
Source: James Bunnell and Roofus

I arrived at the View Park before sunset and set up my cameras near the west end historical plaque. I scanned the horizon with binoculars as darkness gathered, waiting to see if any MLs would make a showing. It soon became dark enough for the brightest stars and planets to begin appearing. A bright light appeared southeast of the View Park but went out before I could take its picture.

A few minutes later a brilliant ML burst into view in the same location. It was the beginning of one of the most spectacular ML events I have recorded. I immediately started a time exposure using a tripod-mounted Pentax film camera. The resulting photograph (Figure 33) would become the cover for my *Night Orbs* book.

Shortly after that first photograph was taken, the ML divided into two MLs and they then performed an interesting dance with each other (see Figure 34). After a short period of back and forth dancing, the right ML started moving rapidly to the west and continued its journey for miles, passing two mercury vapor (MV) lights en route. The left ML continued to oscillate and vary in brightness (pulsing) and exhibited interesting behavior including jets of material that moved down and to the right (see Figure 35). The right ML continued its journey west and soon after passing the MV lights this ML also divided into two MLs. They both continued to journey west for approximately ten miles, then both MLs reversed direction and headed in the direction of the two MV lights they had passed, only to extinguish before reaching them (Figure 36). Meanwhile the original ML (Figure 33) continued to pulse on and off until it also went out for good at 10:11 PM.

Story 18   The Very Next Night
Date: May 8, 2003
Source: James Bunnell and Roofus
Witnesses: James Bunnell from MLVP; Kerr Mitchell from his
               ranch, and a monitoring station at an undisclosed
               location

The very next night another ML appeared and was seen by both Kerr Mitchell, looking east across Mitchell Flat from his ranch house, and by me, as I watched from the View Park. In addition, it was also photographed by one of my monitoring stations. Like the MLs on May 7th, this ML also started from the southeast and journeyed to the northwest. However, the May 8 ML did not reverse after reaching the Presidio railroad tracks; instead, it lost energy and extinguished.

This time I marked the direction at the start of the event and again at its conclusion. These two bearings, combined with bearing information from the monitoring station, made it possible to estimate starting and stopping points for this event. Straight-line distance traveled would have been 11 miles. The initial ML started approximately 13.4 miles from my location at the View Park and approximately 14.2 miles from Kerr Mitchell's observation point. The farthest point reached was over the old Presidio railroad tracks west of Nopal Road, approximately 6.5 miles from the View Park and only 3.4 miles from Kerr Mitchell. Kerr observed that this ML was probably the brightest he had ever seen which is not too surprising given its close proximity to his ranch.

The sighting started at 10:16 PM and ended at 10:34 PM CDT for a total duration of 18 minutes. Figure 37 shows the ML at a height of approximately 250 feet based on comparison to the height of background mesas. There is a large gap in the recorded light-track even though there are no high obstructions to account for such a gap. The gap was caused by the light going out and then coming back on further down track. Similar gap patterns have been observed in many or most ML light tracks. Some of these can be attributed to obstructions but in this case, there are no obstruction that extend that high above the ground. The implication is that something not captured by the photograph continued across gaps while the light was out.

230

Comment:

Notice the explosive expansion at the end of this time exposure (Figure 37). This expansive feature is not characteristic of a mirage and could not have been a mirage given the widely separated views from the View Park and from Kerr Mitchell's ranch. It looks more like a chemical fire that may have obtained a sudden increase in fuel or oxidizer, then expanded and blew itself out in the process (the light did go out at that point). After a short interval the light re-ignited at a lower altitude with a smaller flash and continued its journey to the northwest (Figure 38).

Story 19   Miracle Hunters

Date: July 12, 2003

Source: Personal account: Jim Bunnell plus
                 Discovery Channel program.

Witnesses: Jonathan Levit, Scott Goldie, Jim Linsey, James
                 Bunnell, the camera crew and View Park visitors

In July 2003, I participated in the filming of a documentary for Discovery Channel called "*Miracle Hunters.*" Marfa Lights ended up being a seven-minute segment in this hour-long program that covered a range of unusual things. Naturally the film crews (a private company) were keen to obtain their own video of the Marfa Lights. I kept telling them it was unlikely we would get that opportunity. I had spent too many nights not seeing MLs. We started our hunt deep in Mitchell Flat and drove even deeper. After spending most of the evening on ranch roads without seeing anything unusual, we decided to give it up and head for the motel. However, as we were about to reload into our vehicles, Jim Linsey asked if I would mind going with them to the View Park to obtain video of the host, Jonathan Levit, manning a camera and looking for MLs in the night. "Sure," I said, and off we went to the View Park.

It was a little before midnight on July 12, 2003, when we reached the View Park and started filming. The cameras rolled as Jonathan was asking me to explain how I could tell if we were looking at automobile lights, mercury vapor (MV) ranch lights or ML lights. As if on cue, an unknown light suddenly popped into view roughly south of our location. I recognized right away that this was very probably an ML and moved to my camera to take 35 mm photographs. The camera crew also managed to film this event. After the light went out we discussed what we might have witnessed, and then resumed our planned filming.

In a few minutes the light reappeared and I once more photographed what we were seeing. By then, I had become convinced that this was indeed an ML moving west. The 35 mm pictures I took were sufficient to confirm that observation even though the ML was located too far north and west to be captured by my night monitoring station, Roofus.

We were all delighted to have the good fortune of seeing an ML make two short appearances and to have captured it with the TV crew's professional video equipment. I have no doubt that this was an actual ML because of its location (in close to the View Park where there are no roads), its pulsing on-off behavior, and the large distance between the two appearances. My two time exposures (Figures A6 and A7) look a little different. In the first appearance the light was turning on and off repeatedly, creating a typical dotted light track. In the second appearance, it was on continuously and moved a little faster.

Comment:

The first part of the second photograph shows a small separate light. I have speculated that this might be a camera artifact resulting from my movement of the camera to a new location without waiting for any motion-induced oscillations to dampen out, but there is no way to know if this is an artifact or real.

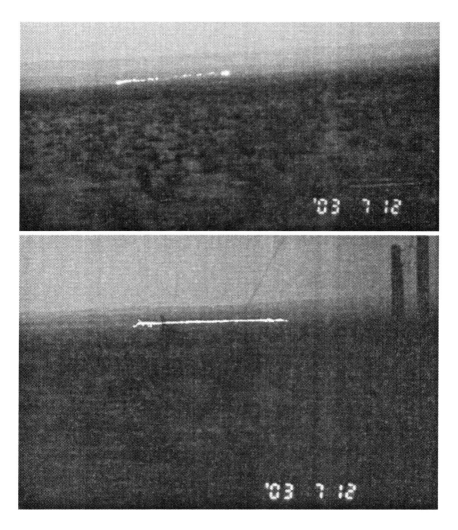

Figures A6 and A7. An ML showed twice while we were shooting film for a Discovery Channel program called "Miracle Hunters."

Story 20   Red MLs and a Green Ball of Light
Date: Late Summer 2003 and early November 2005
Source: Email to editor@nightorbs.net
Witnesses: Dan and wife

Just finished reading your [Night Orbs] book. Very interesting and clarifying.

My wife and I observed [MLs] on our first Marfa visit in 2003. It was late summer, late at night, and we were parked in my pickup looking due east [toward] the low hills just south of Paisano Pass. We witnessed an incredible display of tennis-ball-sized fiery red orbs. We were able to observe these mysterious light balls using Bushnell 16X binoculars. They bounced up in the sky, zoomed together, split apart and did all sorts of crazy things. At times there would be 6-7 of them bouncing around. They would start from fairly far apart, merge into one, split apart again, and then race off in any number of directions. I would guess they were 400-1000 yards south of US 90. It was around one AM and we were the only ones in the MLVP. This [light display] continued for about 30 minutes and then abruptly stopped. We have never seen them again even though I am frequently near Marfa on business and have made numerous evening trips to see the lights.

[In November 2005] we were talking to three couples from Missouri who had rented a van in Alpine and driven over to the MLVP. We were all standing between my pickup and their van facing due south. We were parked left of the shelter on the exit part of the driveway. Without warning and approximately 100 feet directly in front of us, a cantaloupe-sized intensely-green ball fell straight down from the sky and struck the ground. Everyone was startled by this unexpected silent event. It just hit the ground in

234

front of us without sound and went out.

Comment:

The first part of this story is similar to Story 12 with red balls of light exhibiting dynamic-complex movements in the sky but the location (east-southeast of the View Park) is different. The second part, involving a green ball of light that struck the ground near them, is also interesting and different than any other report in this set of stories.

Story 21   Lights Behaving Impossibly
Date: April 9, 2004
Source: Personal account: Jim Bunnell

This story is hard for me to believe in spite of the fact that I am the one who lived it. In April 2004 I was working to develop a new monitoring station known as Snoopy (see previous **Snoopy** discussion). Snoopy is located on a small hill with vulnerability to lightning strikes, so I decided to install a couple of lightning rods. To properly ground them, I drove brass rods deep into the ground using a twelve pound metal fence post driver. At one point I lost control of the hammer and landed it squarely on my left big toe. Ouch!

That night I was tempted to stay in my hotel and nurse my poor toe, but nights in Marfa are precious opportunities to look for MLs and I had located what seemed like a perfect spot to view them from deep within Mitchell Flat. I decided to ignore my painful toe and try out this new viewing location. The snake threat is a special concern at night, so I forced my feet into boots and set off for another night of monitoring.

The new site was accessible only from Highway 67, something that is no longer possible without permission because ranchers have installed locked gates to stop unwelcome interlopers. But ranch roads were open at the time and I drove to my special spot. Before long I wanted to extract my poor toe from that tight boot. I knew if I took off the boot there would be no way to put it back on so I toughed it out until sometime after 10 PM. Able to stand the pain no longer, I packed up cameras and tripods and started driving back to my hotel. The road was rough and I looked forward to the smooth pavement of Highway 67. As I approached Highway 67, a convoy of northbound vehicles drove over a small hill not far from the intersection of the ranch road I was on.

My thought at the time was that I needed to try and get ahead of them so I would not be delayed behind this gaggle of vehicles when we reached the Border Patrol Check Point. I surged ahead and, expertly executing a Texas rolling stop, entered Highway 67 and turned toward Marfa. Expecting that this convoy of vehicles was probably moving 60 or 70 mph, I pushed my accelerator to the floor urging my small V8 to accelerate for all it was worth. At the same time I was watching the oncoming vehicles through my rear view mirrors and was prepared to leave the roadway if necessary to avoid being overrun.

It was immediately obvious that these northbound vehicles were not traveling that fast and in only a moment I realized they had stopped. Easing up on my accelerator, I realized that stopped was not the right word. They were not just stopped, they were moving backwards! With my mouth hanging open in shock, I watched my mirror in amazement as this collection of bright white lights backed up over the hill and out of sight. At no time did I see a red taillight. The number of bright lights would suggest that there

236

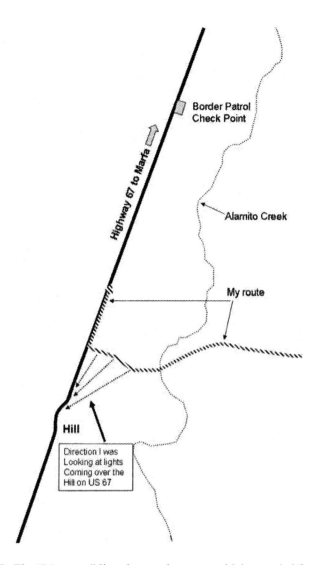

Figure A8. The "My route" line shows where my vehicle traveled from a ranch road onto Highway 67. My view of unknown lights (ULs) that came over a hill in the northbound lane of Highway 67 changed by ~ 40 degrees as I drove forward and entered the highway. These strange lights came over the hill in the northbound lane and then retreated back over the hill and out of sight while remaining in that same lane.

237

were perhaps 6 or 8 large and smaller vehicles. The idea that a convoy of vehicles that large could or would be stupid enough to back rapidly up and over a hill on Highway 67 was unthinkable. The experience was exactly as if I had seen a movie of vehicles driving over the hill, approaching, and then someone reversed the movie and the action ran backwards.

I could not imagine what I had just seen. I was greatly tempted to spin my vehicle around and give chase just so I could find out what in the flying world I had seen. But there was a problem with that plan. I was by then close to the check point and knew for sure that if I chased those lights, the Border Patrol would then chase me, probably insist on searching my truck and I would be delayed for no telling how long. My toe could not stand the thought of that kind of delay. So I decided to go through the check point and then reverse direction.

I had to wait for another vehicle which gave me time to consider that a round robin trip might also look a little suspicious to the BP. By the time I exited the check point those mysterious lights were long gone and my toe was wanting out of those boots, sooner -- not later. So I drove on to my motel thinking on the way about what had just happened.

The only logical explanation would be a mirage. But what kind of mirage would climb a hill and stay exactly in the north-bound lane both coming and going? The idea does not seem plausible because my line of sight changed by 40 degrees as I approached Highway 67 and then remained constant after I turned right onto the highway (see Figure A8). Throughout these observations the lights remained in the northbound lane. This story is so unbelievable that I was tempted to omit it, but have instead told it exactly as it happened.

Story 22   **"Is That One There?"**
Date: May 8, 2004
Source: Personal account
Witnesses: James Bunnell, a bus load of college students and
                 others at the View Park

It was the end of another trip to Marfa without seeing any MLs. I was looking forward to returning home the next day but had time for one more night of peering into the darkness in hopes of seeing something unusual. As I approached Mitchell Flat from Alpine, I could see a large thunderstorm, complete with lightning, unloading rain directly over the spot where I had planned to watch for the night. Knowing it would be wet and muddy where I had planned to set up tripods and cameras, I decided to forego another night of waiting and watching. Instead I would to go to the View Park, talk to a few visitors, then return early to my motel and get rested up for the trip home.

I parked near the east end of the View Park and then entered the shelter carrying only binoculars. I was delighted to find there a group of inquisitive college students; I ended up giving them an impromptu talk on the Marfa Lights. In the middle of my talk, one young lady pointed out into the gathering darkness and asked, "Is that one there?" I raised my binoculars for a look and muttered, "Could be, but it is probably a ranch truck." As I continued with my talk I could see that unknown light moving to the right (north-west) and it had begun to pulse on and off. It then did a step change in brightness, becoming extremely bright. I was almost in disbelief -- the young lady was right. It was indeed an ML! I quickly ended my talk and ran back to my vehicle to extract a tripod, camera and spotting-scope lens. By the time I could get set up, the ML had ended but another one had appeared, also moving northwest into a strong wind. See Figure 39 for the time exposure photograph of this ML. The wind was blowing so hard that I had to hold my 16-pound camera and tripod with one hand for fear that the wind would blow them over. Once my camera completed the picture, I gathered up the entire apparatus and carried it into the View Shelter where I was able to find a wall offering shelter from the wind.

By then, the second ML had also disappeared but soon another burst into view to the southeast, in line with Whirlwind Mesa. It moved down and to the left, then split into multiple lights that danced back and forth and sometimes up and down. Before the show concluded I was able to take three more time exposures (Figures 40, 41 and 42). These images show evidence of continuous spectral profiles on the left and/or right side of each photograph. These profiles were created by a diffraction grating consisting of thousands of small prisms that I had inserted between the five-inch spotting-scope lens and my digital SLR camera.

I was most happy to be returning home from that trip with those prize pictures.

Stories 23   MLs Beyond Mitchell Flat
Date: Exact date unknown
Source: A long-time Marfa rancher

There have been reports that support the idea that MLs are not exclusive to Mitchell Flat. A rancher located in the mountains south of Ranch Road 169 once told me of seeing many MLs from the vantage point of his ranch. His ranch is located at a high elevation, giving him an excellent view of all of Mitchell Flat. One night his car failed on the ranch road about a mile from his house. His only choice was to walk to the house on foot. It was a clear night and as he walked along the road he could see ML activity south of the mesa rim in a location where he knew there were no ranch houses and no roads. It was an experience that once and for all settled for him any question that Marfa Lights do exit.

He discussed what he had observed with a neighbor whose ranch is also south of Ranch Road 169. She lowered her voice and admitted to having seen mysterious lights in the valley where her ranch is located. She said, *"I never talk about them because I am afraid people will think I am crazy."*

I should also add a telephone conversation I had with a rancher located far south of Mitchell Flat near the village of Casa Piedra. In response to my question regarding mystery lights, he replied that he had definitely seen them. He described them as yellow-orange in color and arranged in a vertical stack three deep. "They blinked on and off," he said. This description suggested to me that what he saw may have been a night mirage, a type of ML discussed earlier in this book.

There have also been reports of ML sightings near Fort Davis, in Pinto Canyon, and a few in the Shafter's mountains. Probably the most exciting of the Shafter reports involved a night flight of helicopters as summarized in the next story.

Story 24   Night Vision Only
Date: Told to me in 2004
Source: Story told to me following one of my presentations

This story involves two Border Patrol helicopters making a

direct flight from Presidio to Marfa Municipal Airport. They took off after dark with the lead helicopter flying visually and the second helicopter holding formation with the aid of night vision goggles. Their flight path took them over Shafter, an old mining town located twenty miles north of Presidio and surrounded by high mountains. As they were flying near Shafter, the pilot of the trailing craft saw a ball of light below the lead helicopter. Not knowing what he was seeing, the pilot pulled off his night vision goggles to get a better look but the ball of light disappeared. Able to see nothing unusual with his naked eyes, the pilot pulled his night vision goggles back into place and, presto; there was the ball of light once again. The pilot experimented and could only see the strange light when looking through his night vision goggles. As the two aircraft flew on, the light was lost from view.

Comment:

Is this account real or invented? I believe it did happen but have been unable to locate Border Patrol pilots who are willing to confirm. After hearing this story, I began carrying a night vision device when hunting MLs. I found the device to be very sensitive and helpful for finding dim lights, but in all cases I was able to locate the same light sources with good quality optical aids. I began to use the night vision device to search for any residual content that might otherwise go unnoticed. So far my First Generation night vision device has been unsuccessful in uncovering hidden MLs or ML residuals, but this device is somewhat primitive. The Border Patrol would have been using later, more capable night vision goggles. This story raises questions about the special peculiarities of MLs, as well as being another example of reported ML sightings outside of Mitchell Flat.

Story 25   **A Curtain of Light High into the Night Sky**
Date: Exact date is unknown
Source: Emails with James Nixon III

Looking to the southeast one night on a visit to Marfa, Mr. Nixon was amazed to see a colorful curtain of light extending from just above the horizon to perhaps 12,000 feet. This curtain of light was multicolored, predominantly greenish to rose hues to violet. It appeared during or right after twilight and shimmered for approximately 30 minutes in the gathering darkness.

Comment:

Nixon is a trustworthy witness and not the only one to have seen curtains of light, but this type of mysterious light display is rare. My night cameras on three or more occasions have detected some kind of vertical display above Whirlwind and Mitchell Mesas, an area 20 miles southeast of the View Park. But the events they photographed were of shorter duration, probably not as high, and of unknown color because they are black and white cameras (see Figures 63 and 64).

Story 26   **Electrical MLs (presented on page 97)**
Date: June 2 and 3, 2005
Source: Roofus and Snoopy

Story 27   **A Ranch Light That Wasn't**
Date: August 11, 2006
Source: Personal account

MLs can appear anytime of the night but I usually hold watch only until around midnight. On August 10, 2006, I started packing up a few minutes early and was in my truck driving out by 12 AM CDT. Looking out the side window on my right, I expected to see a greenish mercury vapor (MV) ranch light but saw an orange light instead. Turning my head, I soon located the MV light behind brush and suspected that this new orange light was coming from a seldom occupied ranch house. I anticipated getting a better fix on the location when I climbed out of my truck to open a gate. But on reaching the gate, I noticed that the light had gone out. Too bad. I unlocked the gate, drove through and noticed the light again as I went back to close the gate. As I closed and locked the gate, and the light turned off and then back on again. Oh my gosh! It suddenly dawned on me that this was probably an ML! I had already packed away the cameras in a locked box in my truck, which was now sitting on the wrong side of a locked gate! The best I could do was to fish out a camera and tripod in hopes of getting pictures before the light extinguished for good. I rushed to set up a camera and tripod, then managed to capture several over-the-fence shots of this light. Example photographs are Figures 47 and 48. After the light went out, I looked for any remaining dim light using a night vision device, but found none. This was a night with a full moon so visibility was a little better than usual.

The ML moved left and right. It was located on a bearing of 145 degrees magnetic from my location at the gate, which I estimated to be about ten miles away just below the crest of a mesa notch.

When the light did not reappear for several minutes, I reopened the gate and drove in the direction of its approximate location. By truck, I was only able to get within about five miles of the ML's estimated origin. At that point, I tested with a magnetic hand meter and obtained a reading of 49.1 milligauss approximately one hour after the event ended. This is the highest magnetic deviation I have recorded in connection with an ML event and it served to reinforce the idea that electromagnetic anomalies may well be related to ML appearances. This is not something one would expect if the ML is a mirage or an optical refraction.

Story 28   **"There Were Flames Inside!"**
Date: October 19, 2006
Source: Personal account
Witnesses: James Bunnell and Sandra Dees

In October of 2006 we were watching from monitoring station Owlbert when an ML appeared east of our location around 9 PM CDT. I obtained a series of photographs with my infrared-capable camera. While I was busy tracking and photographing the ML, Sandy had an opportunity to study it through twelve power binoculars. It was the first time she had seen an ML that was optically close. The ML was typical yellow-orange in color but she reported that the center of the ML was reddish and looked like something burning. Her observation is consistent with photographs taken by me on February 19, 2003 during a fairly close encounter at the View Park (see Figures 30, 31, and 32), with Dirk's report of "internal movement" (Story 7) and with the 27-minute ML photographed by Snoopy on July 31, 2008 (Figure 53).

Around 10 PM CDT, the ML returned, moving in the opposite direction. Appearance and photographs were similar (Figures 49, 50, and 51). We would see a third light at 10:35 PM but concluded that it was a ranch truck traveling on Nopal Road.

Story 29   Faint Columns
Date: December 2006 and November 2007
Source: Email submitted to editor@nightorbs.net [edited]
Witnesses: M. Bennett and spouse

Around 7:15 PM, a week before Christmas 2006, we arrived at the View Park and immediately noticed many translucent vertical columns of light that flashed on and off. These faint columns of light were blue in color and extended from the desert floor up high into the sky. Vertical "tubes" of light could be seen all over the place but primarily left of the red flashing telephone tower and closer to Goat Mountain on the left. Spacing between the light columns seemed to be random. Some looked closer to the View Park and some looked further away. They flashed on and off in random order with durations of less than a second each. We witnessed as many as 20 to 30 such light displays and would see them again when we returned to the View Park in November 2007.

Comment:

This report is unique. On a few occasions people at the View Park have told me they were seeing faint vertical columns of light that I could not see. I would discover the cause when it happened to me too. In my case, these were latent images of telephone poles. Staring into the night in hopes of spotting mystery lights, it is sometimes possible to acquire latent images of telephone poles that become imprinted on ones retinas simply because they are in your field of view. Then when you shift your vision to a different direction, those latent images appear as faint-vertical columns of light.

However, that latent image explanation does not work for Bennett's report for a number of reasons:

- They were able to see these columns immediately upon

arrival at the View Park without time for latent images to be imprinted,

    - the reported quantity and random spacing of the columns of light are inconsistent with telephone poles,

    - the blue color is wrong for latent images, and

    - rapid "flashing" of these images is inconsistent with latent images.

Other people have reported seeing columns of light from locations that did not involve nearby telephone poles. However, upon further inquiry, their reports of light columns turned out to be cases of "light stacking." Light stacking is an aspect of night mirages as discussed in Part II. Mr. Bennett reported seeing "tubes" of light, not stacked lights. Also the quantity, distribution, and flashing of these tubes of light are not consistent with the mirage conditions that give the illusion of stacked lights. Bennett's report of faint blue flashing tubes of vertical lights in significant quantities is an important further elaboration of ML and Mitchell Flat uniqueness.

Story 30   **Ball Lightning or MLs?**
Date: August 19, 2007, approximately 9 PM to midnight
Source: Linda Armstrong
Witnesses: Linda Armstrong, her mother Glenda Moseley and Grandmother Verma E. Ward

Linda, her mother and grandmother observed strange lights from the porch of Linda's house east of Marfa, on the grounds of the Luz de Estrella Winery. Starting around 9 PM the three women observed paired balls of light that seemed to be triggered by cloud to cloud lightning strikes. The lightning would flash and then a paired set of reddish light balls would appear and follow an arching flight path up into the sky and back down, followed by a bounce into a second arch. These arch bounces would repeat until the lights went out after several bounces. These strange light displays were repeated a number of times. Near midnight a different kind of light display occurred. Three balls of light seemed to rise out of the ground and shoot high into the sky with incredible speed. Near the top of this display, the light balls seemed to branch out into a widening pattern until they extinguished.

Comment:

The bouncing balls of light could have been ball lightning or MLs but there is no way to know which for certain. Either way, they constituted rare light displays. The final display of three light balls shooting from ground level up high into the sky with expansion near the top is most unusual and reminds me of the giant jet photographed by Roofus and Snoopy in June 2005 (Figure 71), but apparently the event Linda and her family witnessed lasted too long to have been a jet. Her report is interesting and a good illustration of the wide range of light phenomena to be seen east of Marfa in Mitchell Flat.

Story 31	It Moved Up and Down in an Arc
Date:	March 11, 2008, between 9:45 and 10:30 PM
Source:	Email submitted to editor@nightorbs.net
Witnesses:	Don Batory and Family

Don and his family visited the Marfa Lights View Park both Monday and Tuesday. The first night they saw and recognized car lights to the southwest but did not see any lights they considered unusual.

Tuesday night was different. At approximately 9:45, Don and his son were the only ones left on the View Shelter platform. Facing south, an orange-colored light appeared. Don described the light's motion as what you might get if someone held a flash light overhead (12 o'clock position), turned it on, then swung it counter-clockwise down to the 6 o'clock position and turned it off. When the light was first turned on it was dim orange. As it swung downward it became a bright white. At the turnoff position it dimmed to nothing. Without knowing how bright the lights were, it was not possible to estimate how far away they might have been, but Don suspected they were fairly close.

Don and his son would witness more of these light "waves." Sometimes the light would go both up and down as if someone were trying to signal. Soon multiple lights appeared and they all started to drift toward the horizon. Don thinks they may have seen four lights, all dancing or moving about. After watching these displays for approximately 45 minutes the family left for their hotel with the desert light activity still continuing.

Don commented that the ground was never illuminated at any time and that passing car and train lights were all brighter than the unknown lights they were watching in the desert.

After returning home Don experimented to see if he could recreate what they had seen using hand-held flashlights. None of their attempts duplicated the light display they had witnessed and all attempts were easily recognizable as being hand-held flash-lights.

Comment:

This is a credible sighting report from credible witnesses.

249

They saw and recognized automobile lights to the southwest and train lights to the west on both Monday and Tuesday nights. The unusual lights they saw Tuesday night were located further south than is possible for Highway 67 car lights. Based on the up and down light motions described, these could not have been vehicle lights on Nopal Road. The colors, light behaviors, and multiplying lights are all consistent with typical Marfa Mystery Lights. The fact that no light beams were seen would seem to disqualify artificial lights. For that matter, the location of the lights would have been difficult for any would-be hoaxers to reach and certainly uncomfortable, given Marfa's nighttime temperatures in the forties with 6 to 15 mph winds. The fact that surrounding ground was not illuminated suggests that the lights may have been located further away than they appeared.

Story 32   **Chased by a Green Light**
Date: August 12, 2008
Source: Mr. Rob Grotty, Forester for Texas Forest Service

On Tuesday night, August 12, 2008, Mr. Rob Grotty paid a visit to the Marfa Lights View Park and observed unknown lights that seemed to be at a considerable distance from his location. At approximately 10 PM CDT, Rob concluded his observations and headed back to Marfa. As he drove toward Marfa he noticed a following green light in the passenger side mirror of his truck. The light was similar in color to a green traffic light and was not in the roadway behind him. Instead it seemed to be flying above the railroad tracks that parallel US 90. Wondering what this might be, he turned his head to the right and could clearly see this odd light through the untinted back window of his truck. It was a ball of green light that did not radiate or show any evidence of a light beam. Rob was driving 65 mph and this strange light seemed to be about 5 mph faster because it was gaining on him.

The light got within 3 or 4 car lengths of Rob's truck and, curious to know what this light was, he slowed down to allow it to pass. But the light slowed when he slowed. With increasing curiosity, Rob pulled over to the side of the road and stopped. To his amazement the strange green light also stopped. Rob estimated the light to be about ten feet above the railroad tracks and higher than the train section light he could see farther down the tracks. So far as Rob could tell, the light was not attached to anything as it hung suspended in the air above the tracks. Rob turned his truck around to go back for a closer look, but the light disappeared as he was

Figure A9. When Rob pulled over and stopped, the green light also stopped above the railroad tracks that parallel US 90. When he turned his truck around the light disappeared. This is an artist's conception of how the light might have looked if Rob had stepped out of his truck to photograph it.

turning.

Rob drove back in the direction of the View Park looking for anything on the tracks that might account for this strange light. He did not find anything on the tracks, but just before reaching the View Park, he did see a railroad truck parked next to a railroad utility box; the truck was not on the tracks. Rob then resumed his trip to Marfa and the green light did not reappear.

Comment:

Rob Grotty is not the first and I am sure he will not be the last person to encounter strange following lights while driving US 90. The next report occurred at an earlier date but is presented here because of similarities to Rob's account. (See **Part II, Four ML Mirages Stories**, for discussion of potential mirage explanation.)

Story 33   Chased by a White Light
Date: 2004
Source: Ms. Lydia Quiroz, Bank Executive

Ms. Lydia Quiroz of Alpine visited the Marfa Lights observation spot (this occurred before the current View Park was built) along with four other people who also wanted a glimpse of a Marfa Light. It was late when they decided to give up on seeing the lights and return to Alpine. Soon after leaving their roadside viewing location, the driver noticed a single white light closing on them from behind. Being a single light, he assumed it was a motorcycle but it did not radiate or show a beam so he asked his passengers what they thought the light might be. Lydia, sitting in the back seat, looked through the car's rear window and could see a strange but beautiful light approaching in their eastbound lane. As the light came closer, she could see that it was not round but more of an ellipsoid. She described it as a beautiful clear light and, as it came closer, they could see that it was not a motorcycle. It came within a few feet of their car and, at that close distance, was perhaps two and a half times larger than a typical car light would have been.

When the light was almost in contact range, a westbound car

Figure A10. A drawing of the mysterious light that caught up with them and came close.

passed them. The passing car was braking hard and pulling over as it passed, evidently because they had spotted this brilliant light. But at that scary moment the light suddenly disappeared.

Comment:

This fascinating story is similar to the previous story except that the light was white instead of green, was traveling above the roadway instead of above train tracks, and the incident occurred east of the current View Park location instead of west. (See **Part II, Four ML Mirages Stories**, for discussion of potential mirage

253

explanation.)

Story 34   Another Car Chase on US 90
Date: October 8, 2008
Source: Mrs. Linda Armstrong

Mrs. Linda Armstrong left the Holland Hotel in Alpine around 8:45 PM CDT and drove toward her home in Marfa. Soon after passing the Paisano Encampment she noticed a white light following behind her car. The light was large and moving rapidly in her direction. At first she thought it must be a large truck wanting to pass, but as it came closer she could see it was a single bright light. As collision seemed imminent, she involuntarily pressed the accelerator to the floor, but could not outrun the light. She was not sure if the light passed over her car or through it, but it did flood her vehicle with intense white light. Then it was suddenly in front of her vehicle and moving rapidly away. She watched in amazement as it seemed to follow a bend in the road ahead of her. It then left the road and flew through Mitchell Flat south of the View Park.

Upon reaching the View Park Linda leaped from her car and ran to the Marfa Lights viewing platform to grab one set of built-in binoculars that are available there. She watched the receding light while commenting to others at the View Park that they were witnesses to a "real Marfa Light."

Comment:

Dr. Brueske's book of Marfa Light stories includes similar accounts and we also have my strange account, Story 21, involving unknown lights flying above Highway 67. (See **Part II, Four ML Mirages Stories**, for discussion of potential mirage explanation.)

Appendix B

Mystery Light Data

Appendix B
Mystery Light Data

Acquisition of ML data is hindered by two obstacles, (1) infrequency of occurrence and (2) the need to filter out artificial light sources.

I have tackled the first obstacle by developing automatic monitoring stations (Roofus, Snoopy and Owlbert), by making numerous trips to Marfa to collect monitoring station data, and by performing onsite observations as frequently as practical. The combination of these measures has resulted in a small but viable database of 52 documented events on 35 nights (Table B1). Data from these events are summarized in this Appendix.

The second obstacle, filtering out artificial lights, requires extraordinary care because in today's modern world there are many artificial light sources, even in sparsely populated West Texas. As noted earlier, vehicle lights on Highway 67 can be seen nightly from the View Park and they certainly do look mysterious to the uninitiated. Ironically, these lights are easy to filter out because their locations southwest of the View Park are well known and the volume of traffic makes them more than obvious. The same is true of vehicle lights on busy US 90. Ranch lights, tower lights, and lights mounted on structures are all fixed locations, well known, and easy to filter. Train lights are bright and can be seen at great distances, but they travel on rails that do not move, resulting in repeating patterns that are easy to recognize and filter.

Aircraft lights (ACs) of every description, including both fixed wing and rotary wing, fly across Mitchell Flat nightly but they are also easy to recognize and filter. In most cases they employ easy-to-spot rotating beacons used for collision avoidance. Being a retired aerospace engineer, I have no difficulty recognizing aircraft with or without rotating beacons. My monitoring stations are programmed to stack multiple images (usually 32, 64 or 128) before saving to memory. The resultant saved images are time exposures. Aircraft equipped with rotating beacons look like a string of beads in these images but even aircraft flying at night without rotating beacons create distinctive light tracks that are usually easy to recognize. Meteors, sprites, stars, planets, and the moon are also easily recognized and filtered.

The toughest aspect of filtering artificial lights is being able to recognize and filter vehicle lights within Mitchell Flat because this region has many unpaved roads. Most of these are little more than rough trails and rarely used at night, but it does happen. This fact proved problematic for my monitoring stations because early cameras were only able to show lights moving against dark backgrounds, making it difficult to distinguish MLs from vehicle lights.

During the latter part of 2004, I upgraded Roofus and Snoopy with extremely light-sensitive cameras that significantly improved my ability to filter out artificial lights. These improved cameras revealed the likelihood that earlier data were probably contaminated to some degree with vehicle lights. To ensure data purity I elected to discard all earlier designations of mystery lights (MLs) that could not be independently verified. This decision caused valid data to be lost, but was necessary to preserve data integrity.

Thanks to today's supersensitive cameras, I am now able to see more clearly, making it possible to use my own "filter rules" to exclude artificial light sources. Any light that originates at or stops at a known ranch house, originates at a known point of entry (i.e. road entry point), kicks up dust, or that pauses at known locations of ranch gates is automatically assumed to be a headlight. In addition, detection of a light beam automatically labels the light source artificial. Even with these basic rules for filtering out artificial lights, there are still times when available information is inadequate to clearly classify the light source as artificial or mysterious. Questionable light sources are labeled Unknown Lights (ULs) and are kept in a separate database.

Identification and filtering of artificial lights is critical but recognition of ML characteristics is also required before lights can be given ML designations. The ML characteristics I look for are pulsing, on and off states, and step changes in brightness (Splitting and merging, when present, also help establish necessary strangeness). In addition, when I am there to observe, I look for yellow-orange or reddish color and/or sudden changes from yellow-orange to bright red. MLs may exhibit only one or two of these characteristics and sometimes artificial lights can do the same. For example, car lights on Highway 67 may appear to be merging and/or splitting when what the observer is actually seeing is two or more cars following a curved roadway.

In the case of Type IV MLs (i.e., those that fly above the horizon), ML colors tend to be more reddish and reported movements are generally more dynamic than their lower flying Type III brethren. Night cameras do not have the advantage of showing color, so monitoring station images are treated with more caution than is required for personal observations.

Discussion of Selected Data Results

Questions most often asked by individuals planning to visit Marfa are: (1) What are the odds of getting to see mystery lights on any given night, and (2) What are the best times to look for them?

As an ML investigator I would rather omit valid MLs than to risk including artificial lights in saved data. This requirement and the fact that monitoring stations are able to cover only a portion of Mitchell Flat mean that my ML frequency scores are understated. This is especially true prior to 2005 because most monitoring station data collected in 2003 and 2004 were discarded and, prior to 2003, there were no nighttime monitoring stations.

For all these reasons, the chances of an ML appearances on any given night are better than these statistics (Figure B1) would seem to indicate. That, of course, also means that the total number of MLs per year in the Marfa region is unknown, but we do know that such events are rare. Given the extent of nighttime coverage, and my data acceptance criteria, I estimate that my monitoring stations are finding at least half (an average of 9.5 on 5.25 nights) of the MLs that appear in Mitchell Flat, bringing the estimated total to about 19 MLs on 11 nights per year.

Figure B2 shows how soon after sunset MLs appear. Start

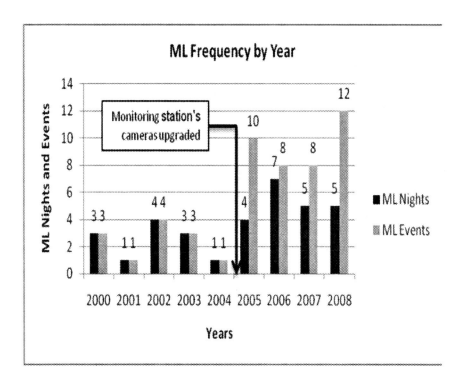

Figure B1. Prior to 2003 there were no monitoring stations and most monitoring station data were discarded as unreliable for 2003 and 2004, resulting in lower ML frequency of appearance for these earlier years. During the period 2005 to 2008, with the help of upgraded monitoring stations, we have identified an average of 9.5 MLs on 5.25 nights per year. Monitoring station and on-site observations are thought to be finding about half of the MLs in Mitchell Flat, making the total annual estimate 19 MLs on 11 nights.

times occur more frequently over the first four hours, accounting for 88% of the MLs recorded. Beyond fours after sunset, ML appearances are less frequent.

Visitors would also like to know if some months are better than others with regard to ML appearances. Are MLs more likely to appear during cold weather or hot weather? During the rainy season? Figure B3 shows data collection by month, although the amount of data available is not yet sufficient to provide solid answers to these good questions. The "best" months to date are

261

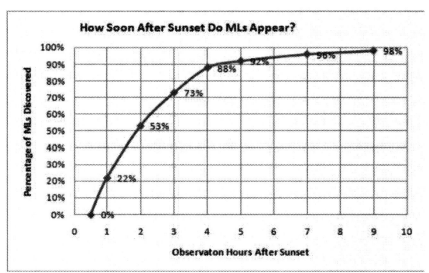

Figure B2. How soon after sunset do MLs appear? These data, from 2001-2008, show first observations occurring no sooner than 1/2 hour after sunset and, on average, 88% of ML appearances occured within the first four hours after sunset.

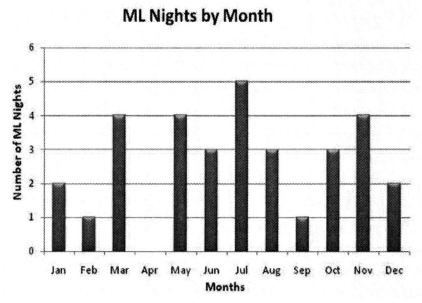

Figure B3. Are some months better than others? March, May, July, and November are the "best" months so far. Absence of MLs in April is most likely due to the small size of this data set.

262

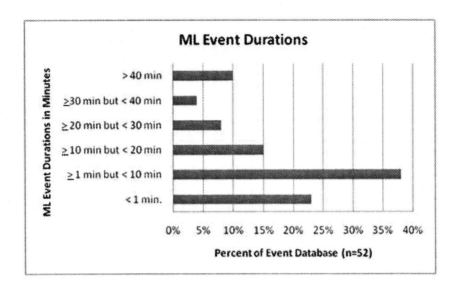

Figure B4. ML durations vary from milliseconds to hours, but most are between 1 and 10 minutes (median = 4 min; n=52).

July, March, May, and November, suggesting that neither rain nor temperature are significant factors. The fact that currently no data is logged for the month of April and only single nights in February and September is believed to result from the small size of this data set.

Another common question is how long do MLs last? Figure B4 shows the distribution of ML durations in minutes. Durations vary from milliseconds to many hours but ninety percent lasted less than 40 minutes with a median time of four minutes. That is long compared to a lightning event, but very little time if you are both trying to figure out what you are seeing as well as attempting to take a picture. It is another reason why there are so few valid photographs of mystery lights. [So remember that when you see an ML, make haste so as not to waste the opportunity. And get your act together before sundown.]

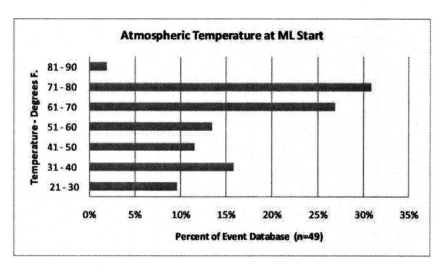

Figure B5. Temperatures at ML start vary from 21 deg. F. to 82 deg. F. (avg. = 59 deg. F; std. dev. = 17 deg. F; median = 63 deg. F; n= 49).

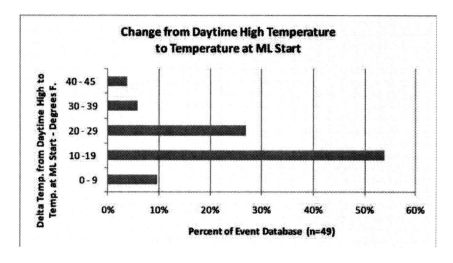

Figure B6. On average, temperature drops 18 deg. F. between daytime high and atmospheric temperature at ML start (std. dev. = 9 deg. F; n=49).

264

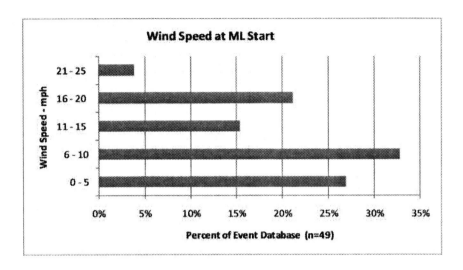

Figure B7. Wind speeds at ML start vary from calm to 25 mph (avg. = 8.4 mph; std. dev. = 5.7 mph; median = 7.5 mph).

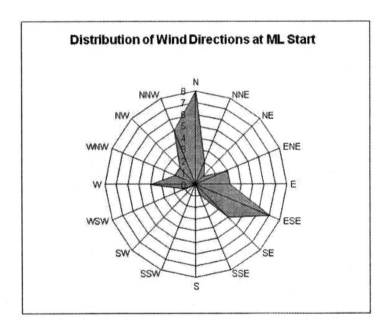

Figure B8. Distribution of wind directions and speeds at ML start are based on wind data collected at the Marfa Municipal Airport. Numbers indicate average wind speeds in mph per direction. Principal directions are from the north and east southeast but these directions may reflect influence of mountains surrounding the Marfa weather station.

265

The preceding charts are based on weather data collected at the Municipal Airport north of Marfa. These data were collected over ten miles from the View Park and probably 15 or 20 miles from the MLs. Even so, in most cases, the data should be reasonably representative of regional weather conditions. These meteorological data do not appear to be related to the appearance of MLs. I also evaluated barometric pressure, humidity and dew point at ML start times, but no relationships were found.

If MLs result from chemical venting out of the ground, then it seems reasonable to expect they will occur in repeating locations. Determination of ML locations requires collection of accurate bearing and timing information from two or more widely separated points of observation. Accurate time information is especially critical in the case of MLs in motion. As part of the 2005 monitoring station upgrades, each monitoring station has been equipped with one or more receivers that are able to collect time data from GPS satellites accurate to within milliseconds. Accurate time data and improved image clarity have enabled computation of ML locations in a dozen cases as shown in Figure B9.

What we are finding is that ML locations are surprisingly varied and, so far, computed locations do not repeat except on subsequent nights. One is tempted to speculate that this information weighs against explanations involving gas and/or liquid venting from beneath the surface because we might expect such vents to be in fixed locations. One possible explanation might be that MLs are products of chemicals embedded in the blanket of tuff that covers all of Mitchell Flat. It could be that liberation of these chemicals exhausts the local supply each time, requiring subsequent chemical liberations to originate elsewhere.

That MLs originate from multiple locations throughout

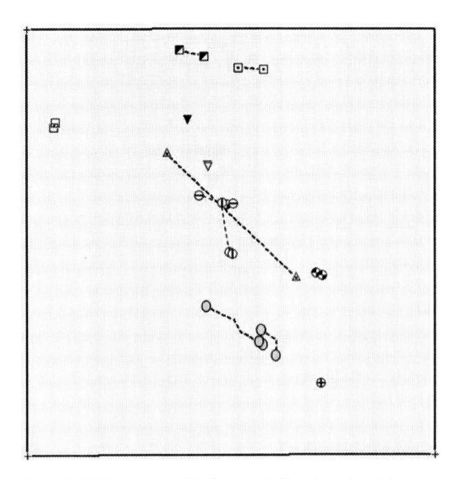

Figure B9. This box measures 25 miles on each side and contains relative location markers for a dozen MLs with known locations. Dotted lines denote ML travel between locations. So far, there have been no repeated locations.

Mitchell Flat is an important clue, but if we take a broad view, it should also be noted that MLs do seem to be aligned and travel in a generally southeast to northwest orientation that parallels major fault lines. There is no way to tell if ML ground tracks are located directly on top of fault lines because available fault line data are imprecise.

I am sure some readers may wish I had included latitude-longitude coordinates of those plotted ML locations. However, to

267

do so might invite an invasion of privately-held ranch land by a few individuals who want to check out these locations for themselves. I have visited as many of these sites as possible and have never found anything unusual or significant. My research would not be possible and you would not be reading this book without the good graces of my rancher friends. Part of my bargain with them is that I will not publish anything that will cause them problems. The important point of Figure B9 is that points of ML origin vary throughout Mitchell Flat.

Mystery Light Data Charts

Table B1 lists each ML event, timing, and related weather data. Events that occurred on the same night as some previous event are enclosed with dotted boundaries.

Table B1, Page 1 of 10. ML Events.

Date (CST/CDT):	11/24/2000	11/25/2000	2/9/2001	1/31/2002
Sunset (CDT/CST)	17:56	17:55	18:38	18:31
Start (CDT/CST)	18:45	19:30	19:30	19:45
Time after sunset	0:49	1:35	0:52	1:14
End (CDT/CST)	24:00:00	11/26/2000 2:00	20:30	20:15
Duration	5:15:00	6:30:00	1:00:00	0:30:00
Sunset (UT)	23:56:00	23:55:00	00:38:00	00:31:00
Date (Universal Time):	11/25/2000	11/26/2000	2/10/2001	2/1/2002
Start (UT)	00:45:00	01:30:00	01:30:00	01:45:00
End (UT)	06:00:00	07:35:00	06:52:00	07:14:00
Moon rise	6:08	7:06	20:33	21:55
Moon set	17:26	18:04	2/10/2001 8:56	2/1/2002 10:03
Was moon up?	No	No	No	No
Max Temp (F) daytime	55	59	84	46
Max Temp (C) daytime	12	15	28	7
Temp at start (F)	44.6	42.8	82.4	32.9
Temp at start (C)	7	6	28	1
Temp Change (day high to ML Start)	10.4	16.2	1.6	13.1
Dew Pt.(F)	17.6	24.8	46.4	4.1
Dew Pt.(C)	-8	-4	8	-15.5
Max Humidy (%)	75%	74%	72%	60%
Start Humidy (%)	34%	49%	28%	30%
Precipitation (in)	0	0	0	0
Precipitation (cm)	0	0	0	0
Sea Level Press (in)	30.12	30.2	30.14	30.31
Start wind direction	WNW	N	SE	N
Start wind speed (mph)	7.5	16.1	5.8	7.5
Start wind speed (km/h)	12.1	25.9	9.3	12
Visibility (miles)	10	10	10	10
Visibility (km)	16	16	16	16

Table B1, Page 2 of 10. ML Events.

3/31/2002	10/30/2002	11/21/2002	2/19/2003	5/7/2003	5/8/2003
19:13	18:11	17:57	18:46	20:36	20:37
19:43	20:55	19:00	20:20	21:55:34	22:15:55
0:30	2:44	1:03	1:34	1:19	1:38
22:43	20:56	19:10	20:30	22:11:23	22:33:35
3:00:00	0:00:10	0:10:00	0:10:00	0:15:49	0:17:40
01:13:00	00:11:00	23:57:00	00:46:00	02:36:00	01:37:00
4/1/2002	10/31/2002	11/22/2002	2/20/2003	5/8/2003	5/9/2003
01:43:00	02:55:50	01:00:00	02:20:00	03:55:34	03:15:55
06:30:00	08:44:50	07:03:00	07:34:00	07:19:34	03:33:35
22:52	1:12	19:16	22:00	5/7/03 11:43	5/9/03 0:44
4/1/2002 8:58	15:09	11/22/2002 8:54	2/20/2003 9:31	5/8/03 1:32	5/9/03 2:19
No	No	No	No	Yes	No
75	64	66	64	89	84
23	17	18	17	31	28
64.4	55.4	28.4	48.2	71.6	62.6
18	13	-2	9	22	17
10.6	8.6	37.6	15.8	17.4	21.4
39.2	35.6	12.2	33.8	14	6.8
4	2	-11	1	-10	-14
93%	81%	51%	max 44%	19%	17%
40%	47%	51%	58%	11%	11%
0	0	0	0	0	0
0	0	0	0	0	0
30.14	30.14	30.45	30.12	29.92	29.88
ENE	N	NW	NE	SE	N
13.8	4.6	8.1	14	4.6	10.4
22.2	7.4	13	24	7.4	16.7
10	10	10	9	10	10
16	16	16	14.5	16	16

Table B1, Page 3 of 10. Dotted line encloses event that occurred on the same night (6/2/2005) as the event previous to it.

Date (CST/CDT):	7/12/2003	5/8/2004	6/2/2005	6/2/2005
Sunset (CDT/CST)	21:00	20:38	20:53	20:53
Start (CDT/CST)	23:15:00	21:04:31	23:38:43	23:38:43
Time after sunset	2:15	0:26	2:45	2:45
End (CDT/CST)	23:30:00	21:08:25	23:38:55	23:42:51
Duration	0:15	0:03:54	0:00:12	0:04:08
Sunset (UT)	02:00:00	01:38:00	01:53:00	01:53:00
Date (Universal Time):	7/13/2003	5/9/2004	6/3/2005	6/3/2005
Start (UT)	04:15:00	02:04:31	04:38:43	04:38:43
End (UT)	04:30:00	02:08:25	04:38:55	04:42:51
Moon rise	7/12/03 20:29	5/9/04 0:25	6/2/05 4:04	6/2/05 4:04
Moon set	7/13/03 5:35	5/9/04 10:27	6/2/05 17:07	6/2/05 17:07
Was moon up?	Yes	No	No	No
Max Temp (F) daytime	91	84	89	89
Max Temp (C) daytime	32	28	31	31
Temp at start (F)	71.6	68.9	75.2	75.2
Temp at start (C)	22	20.5	24	24
Temp Change (day high to ML Start)	19.4	15.1	13.8	13.8
Dew Pt.(F)	48.2	41	60.8	60.8
Dew Pt.(C)	9	5	16	16
Max Humidy (%)	77%	94%	82%	82%
Start Humidy (%)	43%	37%	61%	61%
Precipitation (in)	0	0	0	0
Precipitation (cm)	0	0	0	0
Sea Level Press (in)	30.29	30.14	29.77	29.77
Start wind direction	ESE	E	ESE	ESE
Start wind speed (mph)	8.1	8.1	16.1	16.1
Start wind speed (km/h)	13	13	25.9	25.9
Visibility (miles)	10	10	9	9
Visibility (km)	16	16	14.5	14.5

Table B1, Page 4 of 10. Dotted line encloses ML events that occurred on same night (6/2/2005) as the event previous to it.

6/2/2005	6/2/2005	6/2/2005	6/2/2005	6/2/2005	6/3/2005
20:53	20:53	20:53	20:53	20:53	20:53
23:39:46	23:41:42	23:42:40	23:52:43	23:53:57	0:35:48
2:46	2:48	2:49	2:59	3:00	3:43
23:42:40	23:42:51	23:43:55	23:54:31	0:04:50	3:56:24
0:02:54	0:01:09	0:01:15	0:01:48	0:10:53	3:20:36
01:53:00	01:53:00	01:53:00	01:53:00	01:53:00	01:53:00
6/3/2005	6/3/2005	6/3/2005	6/3/2005	6/3/2005	6/3/2005
04:39:46	04:41:42	04:42:40	04:52:43	04:53:57	05:35:48
04:42:40	04:42:51	04:43:55	04:54:31	05:04:50	08:56:24
6/2/05 4:04	6/2/05 4:04	6/2/05 4:04	6/2/05 4:04	6/2/05 4:04	6/3/05 4:35
6/2/05 17:07	6/2/05 17:07	6/2/05 17:07	6/2/05 17:07	6/2/05 17:07	6/3/05 18:08
No	No	No	No	No	No
89	89	89	89	89	89
31	31	31	31	31	31
75.2	75.2	75.2	73.4	73.4	71.6
24	24	24	23	23	22
13.8	13.8	13.8	15.6	15.6	17.4
60.8	60.8	60.8	60.8	60.8	57.2
16	16	16	16	16	14
82%	82%	82%	82%	82%	88%
61%	61%	61%	65%	65%	60%
0	0	0	0	0	0.33
0	0	0	0	0	0.84
29.77	29.77	29.77	29.79	29.79	29.77
ESE	ESE	ESE	SE	SE	WNW
16.1	16.1	16.1	21.9	21.9	8.1
25.9	25.9	25.9	35.2	35.2	13
9	9	9	7	7	10
14.5	14.5	14.5	11.3	11.3	16

Date (CST/CDT):	11/13/2005	12/16/2005	7/14/2006	7/15/2006
Sunset (CDT/CST)	18:00	17:57	20:59	20:59
Start (CDT/CST)	21:17:43	21:10:34	21:37:21	21:36:45
Time after sunset	3:17	3:13	0:38	0:37
End (CDT/CST)	21:18:15	21:20:14	22:09:04	22:09:04
Duration	0:00:32	0:09:40	0:31:43	0:32:19
Sunset (UT)	00:00:00	23:57:00	01:59:00	01:59:00
Date (Universal Time):	11/14/2005	12/17/2005	7/15/2006	7/16/2006
Start (UT)	03:17:43	03:10:34	02:37:21	02:36:45
End (UT)	03:18:15	03:20:14	03:09:04	03:09:04
Moon rise	11/13/05 16:30	12/16/2005 18:50	7/14/06 23:56	already up
Moon set	11/14/05 4:49	12/17/05 8:54	7/15/06 10:54	1/16/06 12:01
Was moon up?	Yes	Yes	No	Yes
Max Temp (F) daytime	66	50	93	86
Max Temp (C) daytime	18	10	33	30
Temp at start (F)	41	32	78.8	75.2
Temp at start (C)	5	0	26	24
Temp Change (day high to ML Start)	25	18	14.2	10.8
Dew Pt.(F)	24.8	24.8	42.8	48.2
Dew Pt.(C)	-4	-4	6	9
Max Humidy (%)	65%	86%	41%	52%
Start Humidy (%)	53%	75%	28%	38%
Precipitation (in)	0	0	0	0
Precipitation (cm)	0	0	0	0
Sea Level Press (in)	30.27	29.98	30.19	30.27
Start wind direction	NNW	Calm	E	E
Start wind speed (mph)	3.5	Calm	11.5	17.3
Start wind speed (km/h)	5.6	Calm	18.5	27.8
Visibility (miles)	10	10	10	10
Visibility (km)	16	16	16	16

Table B1, Page 6 of 10. Dotted line encloses event that occurred on the same night (10/19/2006) as the event previous to it.

7/16/2006	7/29/2006	8/11/2006	10/19/2006	10/19/2006	12/31/2006
20:59	20:52	20:42	19:21	19:21	18:05
21:47:32	21:43:39	0:02:36	19:36:08	21:05:47	6:28:54
0:48	0:51	3:43	0:15	1:44	12:23
21:49:05	21:50:57	0:07:34	19:57:57	21:08:53	6:28:58
0:01:33	0:07:18	0:04:58	0:21:49	0:03:06	0:00:04
01:59:00	01:52:00	01:42:00	1:21	1:21	00:05:00
7/17/2006	7/30/2006	8/11/2006	10/20/2006	10/20/2006	1/1/2007
02:47:32	02:43:39	05:02:36	1:36:08	3:05:47	12:28:54
02:49:05	02:50:57	05:07:34	1:57:57	3:08:53	12:28:58
7/16/06 0:28	7/29/06 11:09	8/11/06 22:27	10/19/06 5:46	10/19/06 5:46	12/31/06 15:22
7/16/06 13:06	7/29/06 23:29	8/12/06 9:45	10/19/06 18:05	10/19/06 18:05	1/1/07 5:06
No	Yes	No	No	No	No
89	87	91	66	66	53
31	30	32	18	18	11
77	76.1	66.2	55.4	47.3	24.8
25	24.5	19	13	8.5	-4
12	10.9	24.8	10.6	18.7	28.2
46.4	57.2	59	33.8	36.5	21.2
8	14	15	1	2.5	-6
47%	53%	77%	47%	71%	93%
34%	51%	78%	44%	69%	86%
0	0	0	0	0	0
0	0	0	0	0	0
30.22	30.13	30.17	30.1	30.12	30.25
ESE	E	SSE	W	NW	N
5.8	9.8	12.7	10.4	7	4.6
9.3	15.8	20.4	16.7	11.2	7.4
10	10	10	10	10	10
16	16	16	16	16	16

Table B1, Page 7 of 10. Dotted lines enclose events that occurred on the same night as the event previous to it.

Date (CST/CDT):	3/14/2007	6/19/2007	6/19/2007	8/7/2007	8/7/2007
Sunset (CDT/CST)	20:02	21:00	21:00	20:45	20:46
Start (CDT/CST)	0:48:53	21:09:22	21:58:16	22:30:57	22:50:53
Time after sunset	4:46	0:09	0:58	1:45	2:04
End (CDT/CST)	01:13:55	21:12:06	21:59:03	22:32:40	22:52:49
Duration	0:25:02	0:02:44	0:00:47	0:01:43	0:01:56
Sunset (UT)	01:02:00	02:00:00	02:00:00	01:45:00	01:46:00
Date (Universal Time):	3/14/2007	6/20/2007	6/20/2007	8/8/2007	8/8/2007
Start (UT)	5:48:53	2:09:22	2:58:16	3:30:57	3:50:53
End (UT)	6:13:55	2:12:06	2:59:03	3:32:40	3:52:49
Moon rise	3/14/07 5:08	6/20/07 12:24	6/20/07 12:24	8/8/07 2:41	8/8/07 2:41
Moon set	3/14/07 15:21	6/21/07 0:49	6/21/07 0:49	8/8/07 17:42	8/8/07 17:42
Was moon up?	No	Yes	Yes	No	No
Max Temp (F) daytime	80	82	82	72	72
Max Temp (C) daytime	26	27	27	22	22
Temp at start (F)	37.4	63	62	69.8	68.0
Temp at start (C)	3	18	17	21	20
Temp Change (day high to ML Start)	42.6	19	20	2.2	4.0
Dew Pt.(F)	19.4	58	58	62.6	62.6
Dew Pt.(C)	-7	14	14	17	17
Max Humidy (%)	52%	94%	94%	78%	83%
Start Humidy (%)	48%			78%	83%
Precipitation (in)		0	0		
Precipitation (cm)		0	0		
Sea Level Press (in)	30.07	30.20	30.20	30.12	30.12
Start wind direction		N	N	NNW	NNE
Start wind speed (mph)	3.5	9	5	5.8	4.6
Start wind speed (km/h)	5.6	15	8	9.3	7.4
Visibility (miles)	10	10	10	10	10
Visibility (km)	16	16	16	16	16

276

Table B1, Page 8 of 10. Dotted lines enclose events that occurred on the same night as the event previous to it.

9/26/2007	9/26/2007	10/23/2007	1/12/2008	1/12/2008	3/11/2008
19:49	19:49	19:18	18:14	18:14	20:01
21:27:54	23:50:56	4:07:03	21:31:16	23:16:11	21:45:00
1:38	4:01	8:49	3:17	5:02	1:44
21:49:55	23:53:56	04:07:20	21:36:10	23:19:07	22:30:00
0:22:01	0:03:00	0:00:17	0:04:54	0:02:56	0:45:00
00:49:00	00:49:00	01:18:00	00:14:00	00:14:00	02:01:00
9/27/2007	9/27/2007	10/23/2007	1/13/2008	1/13/2008	3/12/2008
2:27:54	4:50:56	10:07:03	03:31:16	05:16:11	03:45:00
2:49:55	4:53:56	10:07:20	03:36:10	05:19:07	04:30:00
9/27/07 20:57	9/27/07 20:57	10/22/07 17:04	1/12/08 10:36	1/12/08 10:36	3/11/08 10:27
9/28/07 9:51	9/28/07 9:51	10/23/07 4:01	1/12/08 22:37	1/12/08 22:37	None
No	Yes	No	Yes	No	Yes
82	82	68	55	55	68
27	27	20	12	12	20
68.0	60.8	24.8	24.8	21.2	43.7
20	16	-4	-4.0	-6.0	6.5
14.0	21.2	43.2	30.2	33.8	24.3
55.4	55.4	12.2	6.8	6.8	31.1
13	13	-11	-14.0	-14.0	-0.5
94%	94%	94%	63%	63%	86%
64%	82%	59%	43%	54%	62%
0	0	0	0	0	0
0	0	0	0	0	0
30.19	30.19	30.45	30.19	30.19	30.32
ENE	NE	NNE	NNW	NNW	North
5.8	4.6	5.8	5.8	8.1	3.5
9.3	7.4	9.3	9.3	13.0	5.6
10	10	10	10.00	10	10
16	16	16	16.00	16	16

Date (CST/CDT):	3/26/2008	3/26/2008	3/26/2008	3/26/2008	3/26/2008
Sunset (CDT/CST)	20:10	20:10	20:10	20:10	20:10
Start (CDT/CST)	21:20:07	21:33:30	21:34:17	21:38:40	22:24:19
Time after sunset	1:10	1:23	1:24	1:28	2:14
End (CDT/CST)	21:33:31	21:34:09	21:35:44	21:39:00	22:31:41
Duration	0:13:24	0:00:39	0:01:27	0:00:20	0:07:22
Sunset (UT)	02:10:00	02:10:00	02:10:00	02:10:00	02:10:00
Date (Universal Time):	3/27/2008	3/27/2008	3/27/2008	3/27/2008	3/27/2008
Start (UT)	03:20:07	03:33:30	03:34:17	03:38:40	04:24:19
End (UT)	03:33:31	03:34:09	03:35:44	03:39:00	04:31:41
Moon rise	3/27/08 1:05	3/27/08 1:05	3/27/08 1:05	3/27/08 1:05	3/27/08 1:05
Moon set	3/27/08 11:06	3/27/08 11:06	3/27/08 11:06	3/27/08 11:06	3/27/08 11:06
Was moon up?	No				
Max Temp (F) daytime	78	78	78	78	78
Max Temp (C) daytime	25	25	25	25	25
Temp at start (F)	55.4	57.2	57.2	57.2	66.2
Temp at start (C)	13.0	14.0	14.0	14.0	19.0
Temp Change (day high to ML Start)	22.6	20.8	20.8	20.8	11.8
Dew Pt.(F)	26.6	26.6	26.6	26.6	21.2
Dew Pt.(C)	-3.0	-3.0	-3.0	-3.0	-6.0
Max Humidy (%)	52%	52%	52%	52%	52%
Start Humidy (%)	33%	31%	31%	31%	18%
Precipitation (in)	0	0	0	0	0
Precipitation (cm)	0	0	0	0	0
Sea Level Press (in)	30.19	30.19	30.19	30.19	30.19
Start wind direction	Calm	Calm	Calm	Calm	W
Start wind speed (mph)	0.0	0.0	0.0	0.0	4.6
Start wind speed (km/h)	0.0	0.0	0.0	0.0	7.4
Visibility (miles)	10	10	10	10	10
Visibility (km)	16	16	16	16	16

Table B1, Page 10 of 10. Dotted lines enclose events that occurred on the same night as the event previous to it.

3/26/2008	5/1/2008	5/1/2008	7/30/2008	8/12/2008	10/8/2008
20:10	20:32	20:32	19:50	20:40	18:32
23:25:30	3:06:29	3:24:49	21:21:10 AM	22:00:00	21:10:00
3:15	6:34	6:52	1:31	1:20	2:38
23:29:27	3:07:02	3:25:17	21:48:48 AM	22:10:00	21:28:40
0:03:57	0:00:33	0:00:28	0:27:38	0:10:00	0:18:40
02:10:00	01:32:00	01:32:00	00:50:00	01:40:00	18:32:03
3/27/2008	5/1/2008	5/1/2008	7/31/2008	8/13/2008	10/9/2008
05:25:30	08:06:29	08:24:49	2:21	03:00:00	02:10:00
05:29:27	08:07:02	08:25:17	2:48	03:10:00	02:28:40
3/27/08 1:05	5/1/08 4:30	5/1/08 4:30	7/30/08 4:01	8/12/08 18:04	10/8/2008 14:56
3/27/08 11:06	5/1/08 16:40	5/1/08 16:40	7/30/08 18:45	8/13/08 3:12	10/9/2008 0:00
	Yes	Yes	No	Yes	Yes
78	82	82	87	84	75
25	27	27	30	28	23
71.6	60.8	59.9	62.6	73.4	60.8
22	16.0	15.5	17.0	23.0	16.0
6.4	21.2	22.1	24.4	10.6	14.2
15.8	12.2	13.0	60.8	62.6	42.8
-9	-11	-11	16	17	6.0
52%	21%	21%	94%	94%	93.0%
12%	15%	16%	88%	69%	52.0%
0	0.0	0.0	0.1	0.0	0.0
0	0.0	0.0	0.1	0.0	0.0
30.2	29.8	29.8	30.1	30.1	30.3
W	W	WSW	WNW	Calm	SSE
11.5	11.5	13.8	19.6	0.0	4.6
18.5	18.5	22.2	31.5	0.0	7.4
10	10	10	10	10	10.0
16	16	16	16	16	16.0

Table B2. For the discriminating observer, GPS coordinates and altitudes for a variety of reference locations in and around Mitchell Flat.

Miscellaneous Locations	Latitude (deg min)	Longitude (deg min)	Altitude (Ft)
East hill US 90 fence post with spike in side	N 30 16.106	W 103 49.205	5064
Ranch fence south side US 90 inline with 1 mile to MLVP sign	N 30 16.468	W 103 51.939	4928
Road center line at old entrance to Marfa AAF	N 30 16.528	W 103 52.810	4931
East picnic table in MLVP	N 30 16.515	W 103 52.902	4922
West picnic table in MLVP	N 30 16.523	W 103 52.976	4913
SE corner of MLVP (fence corner)	N 30 16.475	W 103 52.894	4897
NE corner of MLVP (fence corner)	N 30 16.533	W 103 52.887	4927
SE corner of MLVS	N 30 16.497	W 103 52.963	4898
SE Plaque in MLVP	N 30 16.485	W 103 52.912	4917
NW Plaque in MLVP	N 30 16.513	W 103 53.043	4918
SW Plaque in MLVP	N 30 16.499	W 103 53.067	4911
SW Center Plaque in MLVP	N 30 16.489	W 103 53.012	4913
NW Center Plaque in MLVP	N 30 16.515	W 103 53.003	4914
SE Center Plaque in MLVP	N 30 16.488	W 103 52.987	4916
NE Center Plaque in MLVP	N 30 16.511	W 103 520934	4912
Plaque in front of MLVS	N 30 16.516	W 103 52.962	4917
NW Corner of MLVS wall	N 30 16.516	W 103 52.965	4923
East mid-corner of north MLVS wall	N 30 16.511	W 103 52.973	4919
West mid-corner of north MLVS wall	N 30 16.512	W 103 52.974	4920
SW corner of MLVS wall	N 30 16.505	W 103 52.981	4897
SE corner of MLVS wall	N 30 16.496	W 103 52.962	4922
East mid-corner of south MLVS wall	N 30 16.505	W 103 52.961	4926
West mid-corner of south MLVS wall	N 30 16.506	W 103 52.962	4925
NE corner of MLVS wall	N 30 16.513	W 103 52.960	4923
Center telescope in MLVS	N 30 16.500	W 103 52.971	4928
MLVS metal closet door	N 30 16.508	W 103 52.966	4921
MLVS display case east wall	N 30 16.505	W 103 52.967	4921
MLVS display case west wall	N 30 16.506	W 103 52.969	4922
SW corner of MLVP (fence corner)	N 30 16.489	W 103 53.081	4898
West end of east "D" pavement	N 30 16.540	W 103 52.977	4928
NW corner of MLVP (fence intersection)	N 30 16.547	W 103 53.076	4916
Nopal Road & US 90 (~6 ft beyond cattle guard)	N 30 17.231	W 103 55.435	4793
Nopal Road & railroad crossing	N 30 12.961	W 103 55.163	4695
Power line crossing, south Nopal Road	N 30 10.170	W 103 53.217	4710
Nopal Road locked gate	N 30 9.460	W 103 52.731	4767
US 90 opposite Marfa VORTAC	N 30 17.743	W 103 57.453	4849
Border Patrol Check Point, Highway 67	N 30 15.101	W 104 02.675	4596
Red flashing telephone tower (at gate ~ 15 ft south of tower)	N 30 09.414	W 104 02.480	4393
Center mercury vapor ranch light	N 30 09.106	W 103 51.641	4744
High point on Highway 67	N 30 02.326	W 103 51.631	5417
High mesa MV (not currently burning)	N 30 01.349	W 103 51.384	5193
Center MV light plus 15' pole	N 30 08.256	W 103 51.774	4774

280

Table B3. Directions and distance information from the View Park Center Telescope.

	Mercury Vapor Light (green)		Ranch Light (yellow)		Red Flashing Telephone Tower		High Point On Highway 67	
	Degrees Magnetic	Miles	Degrees Magnetic	Miles	Degrees Magnetic	Miles	Degrees Magnetic	Miles
Center Telescope on Viewing Platform	165	9.6	175	8.9	221	12.6	220	24.8

What About Those Zeolites?

The geological circumstance of a thick layer of zeolitic tuff spread over Mitchell Flat, coupled with the fact that zeolites are not electrically neutral, might be significant factors if MLs involve underground, telluric currents. Zeolites are valued for their absorbency and water is a better electrical conductor than most rock or soil. Water retention in zeolitic tuff might be an important factor if underground electrical conductivity or magnetism are a source of ML displays. My informal tests of samples of tuff from Mitchell Flat showed no magnetic component, but did show the ability of tuff to absorb water. More extensive laboratory and field tests would be necessary to fully explore this avenue of possibilities.

Appendix C

Equipment

Monitoring Station Equipment

My black and white monitoring station cameras are of two types. The Watec 120N (a.k.a. Stellar Cam II) cameras provide controls for image stacking, light intensity, and contrast (gamma). Supercircuit's PC164 cameras are used to extend coverage closer to sunset, sunrise and during a full moon when ambient light is excessive for light sensitive Watec cameras.

Figure C3 illustrates equipment used to obtain accurate time signals from GPS satellites.

Field Equipment

Figures C4 - C12 depict cameras and field meters typically used for field trips. Spectrometer specifications are shown in Table C1.

Figure C1. Stellar Cam II is small but powerful. Yes, that is a quarter sitting on the camera housing.

Figure C2. PC164C Cameras achieve 0.0001 LUX making them useful supplements for twilight periods when the more sensitive Stellar Cam II is in whiteout.

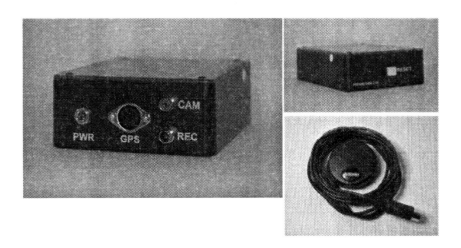

Figure C3. Triangulation of moving lights derived from two or more video streams is not practical unless individual video frames can be accurately matched in time. The above device, known as a KIWI-OSD, is a video time inserter that superimposes a highly accurate time stamp on each video frame. We use four of these devices to enable triangulation of light sources.

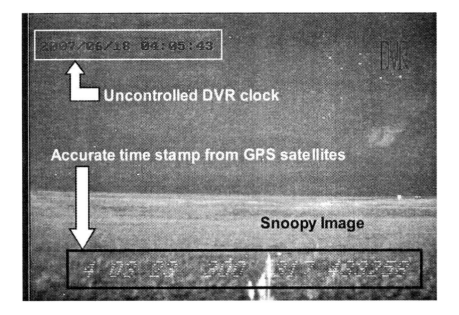

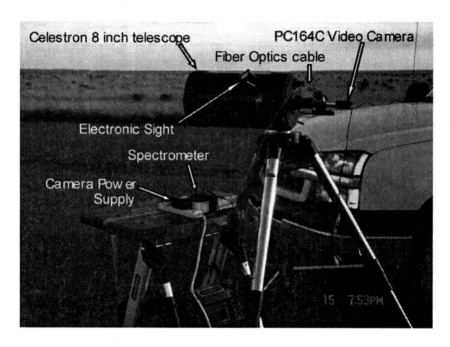

Figure C4. PC Spectrometer in the field. Computer is in the truck.

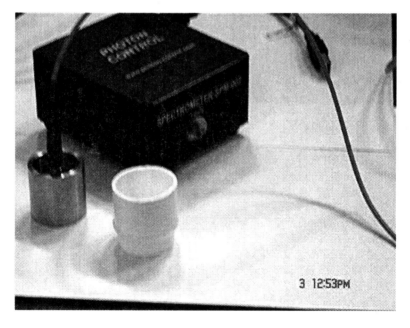

Figure C5. Photon Control SPM002 Compact Spectrometer and puck.

288

Table C1. Spectrometer Specifications

Photon Control Model	SPM-002-A

PERFORMANCE SPECIFICATIONS

Spectral Range (nm)	400-700
Resolution (nm)	0.4 / 0.6
(center / outer spectral range)	
Fiber input	SMA
Focal Length (nm)	70
CCE Pixels	3648
Integration Time Range	10 microseconds to 65 seconds
Triggering	Hardware trigger
Maximum Refresh Rate (Hz)	20

OPERATING SPECIFICATIONS

Temperature (deg. C)	-10 to +55

ELECTRICAL SPECIFICATIONS

Interface	USB 2.0
Power Requirements	10-30 VDC 300 mA Max @ 12 VDC or USB powered

PHYSICAL SPECIFICATIONS

Dimensions (mm)	101.5 x 90 x 53.4
Weight (g)	475
Slit (um)	10

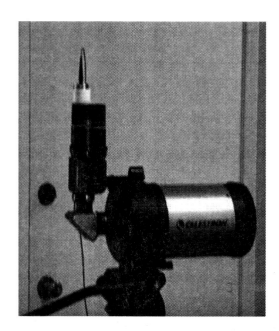

Figure C6. Celestron C5
spotting scope
Schmidt-Cassegrain
Aperture = 127 mm
Focal length = 1250 mm
Focal ratio = 9.84
Field of view = 1 degree
Length = 13 inches
Weight = 96 oz.

Figure C7. Dual mounted 8 inch SCT and C130 Maksutov-Cassegrain with spectrometer, PC-23 and PC-164 video cameras, two 7 inch TVs, 2 mini-digital recorders, and 17 amp power pack.

Figure C8. Air Ion Counter.

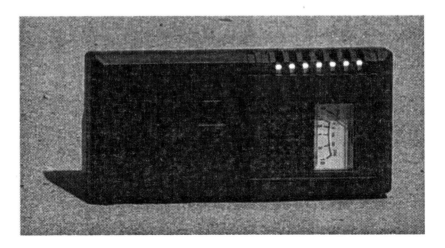

Figure C9. Geiger Counter.

Figure C10. Nikon 7X50 binoculars with built-in illuminated compass are convenient for taking fast and accurate bearings.

Figure C11. Highly sensitive EMF meter with light displays is easy to read in the dark.

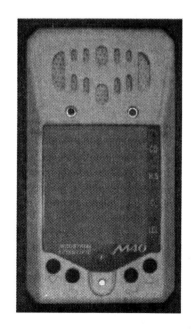

Figure C12. M40 Gas Meter measures presence of combustible hydrocarbons.

Figure C13: EMF meter.

Notes

Please refer to the bibliography for additional references.

1. Mystery light locations include the following references:
   a. http://paranormal.about.com/Library/weekly/ aa080601a.htm *Spooklights: Where to Find Them.*
   b. www.ghosts.org/ghostlights/ghostlights.html *Obiwan's UFO-Free Paranormal Page > Ghostlights*
   c. Corliss, William R. *Lightning, Auroras, Nocturnal Lights, and Related Phenomena.* The Sourcebook Project, Glen Arm, P.O. Box 107, MD21057. 1982.
2. Miles, Elton. *Tales of the Big Bend.* seventh printing, 1999, Texas A&M University Press, College Station, TX 1976
3. Kutz, Jack. *Mysteries & Miracles of Texas.* Rhombus Publishing Company, Inc., P.O. Box 806, Corrales, NM 78048. 1994
4. MacLeod, William. *Big Bend Vistas, A Geological Exploration of the Big Bend.* Texas Geological Press, P.O. Box 967, Alpine, TX 79831. 2002. www.bigbendvistas.com
   Spearing, Darwin. *Roadside Geology of Texas.* seventh printing, May 2001, Mountain Press Publishing Company, P.O. Box 2399, Missoula, MT 59806, (406) 728-1900. 1991.

McGookey, Donald P. *Geologic Wonders of West Texas.* 203 W. Wall, Suite 705, Midland, Texas 79701. 2004.

5. Discussion of Fata Morgana Mirage conditions include the following references:

 a. http://jackstephensimages.com/Merchant/ photographicgallery/fatamorgana/title.html *Fata Morgana.*

 b. www.wikipedia.org/wiki/Fata_Morgana *Fata Morgana.*

 c. www.abc.net.au/science/news/stories/s818193.htm, *Mystery of the Min Min Lights explained.* 28 March 2003

6. www.oulu.fi/~spaceweb/textbook/radbelts.html. *Radiation (Van Allen) belts.* 21 December 1998.

 Russell, C. T. *The Magnetosphere.* Institute of Geophysics and Planetary Physics (IGPP) and Department of Earth and Space Science (ESS), University of California, Los Angeles, CA, Copyright by Terra Scientific Publishing Company, Tokyo. 1987.

7. Walt, Martin. *Introduction to Geomagnetically Trapped Particles.* Cambridge Atmospheric and Space Science Series. 1994.

 The equation used (3.18): $r = R_0 \cos^2 X$ where $r =$ spherical polar coordinate Earth centered, $R_0 =$ Distance from the center of the Earth to the equatorial crossing point of a magnetic field line, $\cos^2 =$ cosine squared, and $X =$ geometric latitude.

8. www.oma.be/BIRA-IASB/Public/Research/Belts/ Particles.en.html *Energetic particles of the magnetosphere,* M. Kruglanski

 www-istp.gsfc.nasa.gov/Education/wradbelt.html *The Radiation Belts,* Dr. David P. Stern. 2002.

www-istp.gsfc.nasa.gov/Education/wtrap1.html, *#9.
Trapped Radiation.*

9. Robb, Rita. *UFO Enigma Chat Log, Destination Space.*
    Chat Date: 10/29/2000.

10. http://en.wikipedia.org/wiki/Pyrophoricity
    http://explanation-guide.info/meaning/Pyrophoricity.html

11. Characteristics of Ball Lightning include the following
    references:

    a. www.amasci.com/tesla/bltalb.txt, Subject: Re: *Ball
    Lightning - new info From:
    bo964@FreeNet.Carleton.CA (Michel T.
    Talbot)* Date: 1995/04/22 Newsgroups:
    sci.geo.meteorology

    b. www.sciam.com/
    print_version.cfm?articleID=000CC3F0-66E4-
    1C71-9EB7809EC588F2D7 Scientific
    American.com, *Periodically I hear stories
    about ball lightning. Does this phenomenon
    really exist?*

12. *The Earth's Electrical Environment* by Geophysics Study
    Committee, Geophysics Research Forum, Commission on
    Physical Sciences, Mathematics and Resources, and
    National Research Council. ISBN: 978-0-309-03680

13. Caglar, I.; Eryildiz, C. *Monitoring of the Telluric Currents
    Originated by Solar Related Atmospheric Events in
    Northwestern Turkey.* 1991

Bibliography

Bord, Janet and Colin. *Unexplained Mysteries of the 20th Century.* Chicago: Contemporary Books, Inc., 1989.

Brueske, Judith M. *The Marfa Lights:* second revised edition. Alpine: Ocotillo Enterprises, 1989.

Bunnell, James. *Seeing Marfa Lights*: *A View's Guide.* Cedar Creek: Lacey Publishing Co., 2001.

---. *Night Orbs*. Lacey Publishing Co., Cedar Creek, 2003.

Cartwright, Gary. *The Marfa Lights.* Austin: Texas Monthly, 1984.

Corliss, William R. *Lightning, Auroras, Nocturnal Lights, and Related Phenomena.* Glen Arm: The Sourcebook Project, 1982.

---. *Remarkable Luminous Phenomena In Nature: A Catalog of Geophysical Anomalies.* Glen Arm: The Sourcebook Project, 2001.

Dash, Mike. *The Ultimate Exploration of the Unknown Borderlands.* New York: The Overlook Press, 2000.

Douglas, Gene. Pecos: *Pecos Enterprises,* 1976.

Eby-Martin, Sharon. *The Mysterious Mirage Effect of the Marfa Lights.* Marfa: Paper on file in the Marfa Public Library, 2001.

---. *We Saw The Marfa Lights!* Marfa: Paper on file in the Marfa Public Library, 2000.

Frohlich, Cliff, and Davis, Scott D. *Texas Earthquakes.* Austin: UT Press, 2002.

Graham, Jeff M. *Letters to the Editor.* Alpine: The Avalanche, 1976.

Hanners, David. *Marfa Lights Convince Pair of Geologists.* Dallas: The Dallas Morning News, 1982.

Hendricks, Edson. *Can Science See the Marfa Lights?* Marfa: Paper on file in the Marfa Public Library, 1991.

---. *Marfa Lights Speculations.* San Diego: 2003.

Jasinski, Laurie E. and Olson, Donald W. Austin: *Texas Highways,* 2000.

Justice, Glenn. *Little Known History of the Texas Big Bend.* Odessa: Rimrock Press, 2001.

Kaczmarek, Dale. *Illuminating the Darkness, The Mystery of Spook Lights.* Oak Lawn, IL: A Ghost Research Society Press Publication, 2003.

Kozicka, Maureen. *The Mystery of the Min Min Light.* Cairns, Australia: Bolton Inprint, 1994.

Kutz, Jack. *Mysteries & Miracles of Texas.* Correlas, NM: Rhombus Publishing, 1994.

Lapedes, Daniel N. *Dictionary of Scientific and Technical Terms.* New York: McGraw-Hill Book Company, 1974.

McGookey, Donald P. *Geologic Wonders of West Texas.* Midland: McGookey, 2004.

Miles, Elton. *Tales of the Big Bend.* Seventh printing. College Station: Texas A&M University Press, 1999.

---. *More Tales of the Big Bend.* College Station: Texas A&M University Press, 1988.

MacLeod, William. *Davis Mountain Vistas, A Geological Exploration of the Davis Mountains.* Alpine: Texas Geological Press, 2005.

---. *Big Bend Vistas, A Geological Exploration of the Big Bend.* Alpine: Texas Geological Press, 2002.

Nelson, Robert. *Marfa Lights May Hold Future Energy promise.* El Paso: El Paso Times, 1993.

Pfister, Nancy J. *Marfa Lights Still Tickle Viewers Imagination.* El Paso: El Paso Times, 1990.

Riggs, Rob. *In the Big Thicket, On the Trail of the Wild Man, Exploring Nature's Mysterious Dimension.* New York: ParaView Press, 2001.

Robinson, Jeanne Marie. *Alienated.* Murfreesboro: Greenleaf Publications, 1997.

Russell, C.T. *The Magnetosphere.* Tokyo: Institute of Geophysics and Planetary Physics (IGPP) and Department of Earth and Space Science (ESS), 1987.

Teodorani, Massimo. *A Long-Term Scientific Survey of the Hessdalen Phenomenon,* Vol. 18, No. 2, *Journal of Scientific Exploration.* Lawrence: A Publication of the Society for Scientific Exploration, 2004.

Thornsburg, William K. *The Marfa Lights, A Close Encounter.* Victoria, Texas: 2007.

Treat, Wesley, Shade, Heather, and Riggs, Rob. *Weird Texas, Your Travel Guide to Texas' Local Legends and Best Kept Secrets.* New York: Sterling Publishing Company, 2005.

Tyler, Ronnie C. *The Big Bend, A History of the Last Texas Frontier.* Washington D.C.: Office of Publications, National Park Service, 1975.

Syers, Ed. *Ghost Stories of Texas.* Waco: Texian Press, Doubleday Dell Publishing Group, Inc., 1981.

Spearing, Darwin. *Roadside Geology of Texas.* Missoula: Mountain Press Publishing Company, 1991.

Walt, Martin. *Introduction to Geomagnetically Trapped Radiation.* New York: Press Syndicate of the University of Cambridge, 1994.

Warnock, Kirby F. *Ghost Lights.* Dallas: Big Bend Quarterly, 2000.

Wright, Elwood, and Kenney, Pat. *The Enigma Lights of Marfa, An Unexplained Phenomenon.* Marfa: Paper on file in the Marfa Public Library, 1973.

Glossary

Most of the following definitions are from *Dictionary of Scientific and Technical Terms* published by McGraw-Hill Book Company, Daniel N. Lapedes, Editor in Chief, 1974, and from Wikipedia or other internet sources.

Aurora

A colorful visual phenomenon that periodically occurs mainly at high latitudes in the night sky. Auroras are a result of collisions between atmospheric gases and precipitating charged solar particles (mostly free electrons). Aurora altitudes are typically 100 to 250 km above the ground. Auroras in the Northern Hemisphere are called Aurora Borealis or "Northern Lights." Auroras in the Southern Hemisphere are called Aurora Australis.

Bow shock

Where supersonic solar wind is slowed to subsonic speeds as a result of encountering the Earth's magnetosphere.

Collimator

A collimator is a device that narrows a beam of particles or waves. To "narrow" can mean either to cause the directions of motion to become more aligned in a specific direction (i.e., collimated or parallel) or to cause the spatial cross section of the beam to become smaller.

Coronal mass ejection (CME)

Outflow of plasma ejected from the sun by solar eruptions, filaments or flares. When Earth directed, CMEs can cause shock waves to impinge on the Earth's magnetosphere.

Cosmic ray

Electromagnetic radiation of extremely high frequency and energy; emitted by stars located beyond our MilkyWay galaxy. Collisions between cosmic rays and atoms from the Earth's upper atmosphere result in creation of plasma, some of which becomes trapped in the Earth's inner radiation belt.

Diffraction Grating

Diffraction grating is an optical device consisting of clear plastic or glass material that has been grooved to create thousands of tiny prisms that parse light rays into constituent spectra.

Dipole

Any object or system that is oppositely charged at two points, or poles.

eV

Electron Volts.

Fata Morgana

A complex mirage characterized by multiple distortions of images, generally in the vertical, so that objects such as cliffs or houses are distorted and magnified into, for example, fantastic castles. Fata Morgana mirages are common in Mitchell Flat and can transport and distort lights so they appear where they are not and in shapes that they are not.

Flux

Rate of flow of a physical quantity through a reference surface.

Geomagnetic field

The magnetic field in and around the Earth. Intensity of the field at the Earth's surface is approximately 32,000 nT* at the equator and 62,000 nT at the North Pole. The geomagnetic field can be approximated by a dipole field inclined to the Earth's rotational axis by about 11.5 degrees but this can change as magnetic north is dynamic and drifts over time.

*Tesla is a unit of magnetic induction equal to one Weber per square meter. A nano-Tesla (nT) is one billionth of a Tesla.

Hypergolic

Chemicals that are capable of igniting spontaneously upon contact.

Interplanetary Magnetic Field (IMF)

The sun's magnetic field, which is carried by the solar wind. It impinges on the Earth's magnetosphere and the angle of impingement varies over time because of the sun's rotation.

Ion

An isolated electron or positron or an atom or molecule which by loss or gain of electron(s) has acquired a net electrical charge. An ion can be an electron but common convention adhered to in this book is to use the term ion when referring to free protons (i.e. positrons).

Ionosphere

Region of the Earth's upper atmosphere containing free electrons and ions produced by ionization of the constituents of the atmosphere by energetic particles and by ultraviolet radiation at very short wavelengths.

keV

Thousands of electron volts

Magnetic bottle

Another term for magnetosphere

Magnetic field line

Lines used to represent the magnetic induction in a magnetic field so that they are parallel to the magnetic induction at each point, and so that the number of lines per unit area of a surface perpendicular to the induction is equal to the induction.

Magnetometer

An instrument for measuring the magnitude and sometimes also the direction of a magnetic field.

Magnetopause

The boundary between the solar wind and the magnetosphere where the pressure of the magnetic field equals the dynamic pressure of the solar wind.

Magnetosphere

The magnetic cavity surrounding the Earth that is carved out of the passing solar wind.

Magnetotail

Extension of the magnetosphere on the shadow side of the Earth caused by interaction with the solar wind. True dimensions of the magnetotail vary with solar pressure but are estimated to be on the order of 1000 times the Earth's radius.

meV

Millions of electron volts

Mirror Points

The point where the spiral pitch angle of trapped particles exceeds 90 degrees causing the spiraling particles to reverse direction. Mirror points are located approximately sixty miles above the Earth's surface (see Pitch angle).

Mitchell Flat

A local name given the desert region located about nine miles east of Marfa, Texas where the Marfa Lights View Park is located and where mystery lights (MLs) are most frequently seen.

ML or ML(s)

Mystery Light(s).

MLMS1

Mystery Light Monitoring Station One (Roofus)

MLMS2

Mystery Light Monitoring Station Two (Snoopy)

Particle

Any very small part of matter, such as a molecule, atom, proton, ion, or electron.

Pitch angle (of the helix)

The angle between the particle velocity and the magnetic field and is defined as: pitch angle = arc tan^1 (perpendicular velocity component / parallel velocity component) [reference **Introduction to Geomagnetically Trapped Radiation** by Martin Walt]

Plasma

A completely ionized gas composed entirely of nearly equal numbers of positive and negative free charges (i.e. free positive ions and free negative electrons). Plasma is a fourth state of matter and the most common form of matter in the universe because stars are plasma.

Pyrophoricity

A pyrophoric substance ignites spontaneously when it comes into contact with atmospheric oxygen. This autoignition capability may depend on substance configuration. Examples would include iron sulfide and many reactive metals when powdered or sliced thinly. Pyrophoric substances can be solids, liquids or gases. Pyrophoric materials are often water reactive as well and will ignite when they contact water or humid air.

Radiation belts

There are two radiation belts wrapped around the Earth. They are known as the inner and outer belts and also as the Inner and Outer Van Allen Belts. The inner belt is part of the plasma sphere and co-rotates with the Earth. Maximum proton density of the inner belt is at an altitude of approximately 5000 km. Inner belt protons are mostly high-energy particles in the million electron volt range that are the products of collisions between cosmic rays from the galaxy and upper atmospheric particles.

The outer belt extends to the magnetopause (limit of the magnetosphere) on the sunward side and to about six times the Earth's radius on the night side. The altitude of maximum proton density is near 16,000-20,000 km. Outer belt protons have lower energy on the order of 200eV to 1MeV and come from the solar wind.

Temperature inversion

A layer in the atmosphere in which temperature increases with altitude. These inversion layers sometimes act like lenses bending light and producing mirages.

Telluric currents

Currents flowing through the ground due to natural causes, such as the Earth's magnetic field or auroral activity. These underground currents are created by the Earth spinning in a magnetic field and are influenced by interactions between the solar wind and the magnetosphere.

Universal Time (UT)

Universal Time, also known as Greenwich Mean Time (GMT) and Zulu Time (Z) is mean solar time at the Meridian in Greenwich, United Kingdom. It is six hours ahead of Central Standard Time (CST) and five hours ahead of Central Daylight Time (CDT). My monitoring stations sample this time reference from GPS satellites.

Van Allen Belts

See radiation belts

Zeolite

A group of white or gray (at least in Mitchell Flat) hydrous tectosilicate minerals characterized by an aluminosilicate tethahedral framework, ion-exchangeable large cations, and loosely held water molecules permitting reversible dehydration. These minerals have unique absorption properties and they are abundant in Mitchell Flat.

Abbreviations and Acronyms

AAF	Army Air Field
AC	Aircraft Lights
AL	Automobile Lights
ATV	All Terrain Vehicle
BL	Ball Lightning
BP	Border Patrol
B&W	Black & White
C	Celsius
CDT	Central Daylight Time
CE	Chemical-Electromagnetic
CST	Central Standard Time
CME	Coronal Mass Ejection
CMV	Center Mercury Vapor Light
Ctr	Center
deg	degrees
dev	deviation
DF	Diffraction Grating
DVR	Digital Video Recorder
EM	Electromagnetic
EMF	Electromagnetic Field
EMV	East Mercury Vapor Light
eV	electron volts

F	Figure
F	Fahrenheit
Hwy	Highway
IR	Infrared
km	kilometers
L	left
LC	Light Curtain
M	Magnetic
M	Mirage
meV	million electron volts
ML	Mystery Light
MLMS1	ML Monitor Station One (Roofus)
MLMS2	ML Monitor Station Two (Snoopy)
MLVP	Marfa Lights View Park
MLVS	Marfa Lights View Shelter
MML	Marfa Mystery Lights
mph	miles per hour
MV	Mercury Vapor Light
NASA	National Aviation and Space Agency
NOAA	National Oceanic and Atmospheric Agency
Owlbert	Name of monitoring station three
PC	Personal Computer

PE	Personal Experience
R	right
RL	Ranch Light
Roofus	Name of monitoring station one
RV	Recreational Vehicle
S	Story
sec	seconds
SG	Sky Glows
SLR	Single Lens Reflex
Snoopy	Name of monitoring station two
std	standard
SUV	Sport Utility Vehicle
UFO	Unidentified Flying Object
UL	Unknown Light
US	United States
UT	Universal Time
VL	Vehicle Light
WS	Weird Stuff

Index

315

About the Author

James Bunnell grew up in West Texas, graduated from Marfa High School and received his engineering degree from New Mexico State University. Graduate studies in Psychology and Aviation Systems followed at Geogia State University and University of Tennessee Space Institute. He was a member of the launch team for all manned Apollo launches (1968-1973) and earned an Apollo Achievement Award for support of the Apollo 11 launch. He later served as project manager for solid rocket motor testing (Minute Man, Poseidon, and Polaris missiles) at Arnold Research in Tennessee where he received a commendation for his study of propellant detonations. He has also served as program manager for various software projects with McDonnell-Douglas, General Dynamics, GDE Systems, Tracor and BAE Systems. Jim retired from BAE Systems as Director of Mission Solutions for Air Force programs and now resides in Fort Worth. He has written two other books: *Seeing Marfa Lights* and *Night Orbs*.